PUBLIC HEALTH CONCERN

SMOKING, ALCOHOL AND SUBSTANCE USE

HEALTH AND HUMAN DEVELOPMENT
JOAV MERRICK - SERIES EDITOR –
NATIONAL INSTITUTE OF CHILD HEALTH AND HUMAN DEVELOPMENT,
MINISTRY OF SOCIAL AFFAIRS, JERUSALEM, ISRAEL

Adolescent Behavior Research:
International Perspectives
Joav Merrick and Hatim A Omar
(Editors)
2007. ISBN: 1-60021-649-8

Complementary Medicine
Systems: Comparison and
Integration
Karl W Kratky
2008. ISBN: 978-1-60456-475-4

Pain in Children and Youth
P Schofield and J Merrick (Editors)
2008. ISBN: 978-1-60456-951-3

Obesity and Adolescence:
A Public Health Concern
Hatim A Omar, Donald E Greydanus,
Dilip R Patel and Joav Merrick
(Editors)
2009. ISBN: 978-1-60692-821-9

Poverty and Children:
A Public Health Concern
Alexis Lieberman
and Joav Merrick (Editors)
2009. ISBN: 978-1-60741-140-6

Living on the Edge: The
Mythical, Spiritual, and
Philosophical Roots of Social
Marginality
Joseph Goodbread
2009. ISBN: 978-1-60741-162-8
(Hardcover)
2011. ISBN: 978-1-61470-192-7
(E-book)
2013. ISBN: 978-1-61122-986-8
(Softcover)

Social and Cultural Psychiatry
Experience from the Caribbean
Region
Hari D Maharajh and Joav Merrick
(Editors)
2011. ISBN: 978-1-61668-506-5

Challenges in Adolescent Health:
An Australian Perspective
David Bennett, Susan Towns and
Elizabeth Elliott and Joav Merrick
(Editors)
2009. ISBN: 978-1-60741-616-6

Children and Pain
P Schofield and J Merrick (Editors)
2009. ISBN: 978-1-60876-020-6

PUBLIC HEALTH CONCERN

SMOKING, ALCOHOL AND SUBSTANCE USE

JOAV MERRICK, M.D.

AND

ARIEL TENENBAUM, M.D.

EDITORS

New York

Library of Congress Cataloging-in-Publication Data

Public health concern : smoking, alcohol and substance use / editors, Joav Merrick and Ariel Tenenbaum.
    pages cm
  Includes index.
  ISBN 978-1-62948-424-2 (hardcover)
  1. Smoking--Health aspects. 2. Drinking of alcoholic beverages--Health aspects. 3. Substance abuse--Health aspects. 4. Public health. I. Merrick, Joav, 1950- II. Tenenbaum, Ariel.
  RA645.T62.P83 2013
  362.29'6--dc23
                          2013039441

Published by Nova Science Publishers, Inc. † New York

Contents

Introduction   **1**

Chapter I   Smoking, alcohol and substance use   **3**
                  Ariel Tenenbaum and Joav Merrick

Section 1: Smoking   **7**

Chapter II   Smoking among adolescents living in the West Bank
                  (Judea and Samaria)   **9**
                  Emmanuel Rudatsikira, Adamson S Muula
                  and Seter Siziya

Chapter III   Smokeless tobacco use among university students   **21**
                  Mohammad Ali Emamhadi, Maryam Jalilvand,
                  Zahra Nikmanesh and Yahya Kazemi

Chapter IV   Alcohol use, cigarette use, delinquency
                  and Thai parenting practices   **33**
                  Brenda A Miller, Hilary F Byrnes,
                  Pamela K Cupp, Aphichat Chamratrithirong,
                  Orratai Rhucharoenpornpanich,
                  Warunee Fongkaew, Michael J Rosati,
                  Warunee Chookhare and Rick S Zimmerman

Chapter V   Smoking from early to late adolescence in Greece   **53**
                  Evdoxia Kosmidou, Mary Hassandra,
                  Antonis Hatzigeorgiadis, Marios Goudas
                  and Yiannis Theodorakis

Contents

Chapter VI   Sleep duration, smoking, alcohol and sleep quality   **67**
Bruce Kirkcaldy and Timo Partonen

Chapter VII   Cigarette smoking among Nigerian
adolescent school boys   **85**
*Alphonsus N Onyiriuka
and Rita C Onyiriuka*

Chapter VIII   Smoking, smoking patterns and perceived
stress in Sri Lankan undergraduates   **97**
*Bilesha Perera, Mohammad R Torabi,
Chandramali Jayawardana
and Ramani Perera*

Section 2: Alcohol   **111**

Chapter IX   Predicting alcohol consumption among
Chilean youth   **113**
*Yoonsun Han, Andrew Grogan-Kaylor,
Jorge Delva and Marcela Castillo*

Chapter X   Alcohol and drug screening of occupational
drivers for preventing injuries   **131**
Robin Christian

Section 3: Other substance use   **135**

Chapter XI   Transitional drug use: Switching from alcohol
disability to marijuana creativity   **137**
*Hari D Maharajh, Jameela K Ali
and Mala Maharaj*

Chapter XII   Substance use and the workplace: Adolescent
and young adult employees   **149**
*Jessica Samuolis, Kenneth W Griffin,
Christopher Williams, Brian Cesario
and Gilbert J Botvin*

Section 4: Interventions   **161**

Chapter XIII   Decreasing student tendency towards smoking   **163**
*Mohammad Ali Emamhadi, Farideh Khodabandeh,
Maryam Jalilvand, Mina Hadian
and Gholam Reza Heydari*

Chapter XIV   Spiritually-based activities as a deterrent to
substance abuse behavior   **173**
Patricia Coccoma and Scott Anstadt

Chapter XV   Buprenorphine for the management
of opioid withdrawal   **189**
Emily Haesler

Chapter XVI   Legislative smoking bans for reducing secondhand
smoke exposure, smoking prevalence
and tobacco consumption   **195**
Janice Christie

Section 5: Acknowledgments   **201**

Chapter XVII   About the editors   **203**

Chapter XVIII   About the National Institute of Child Health
and Human Development in Israel   **205**

Chapter XIX   About the book series
"Health and Human Development"   **209**

Section 6: Index   **215**

Index   **217**

Introduction

Smoking has well-known health hazards. It is associated with lung cancer, oral cancer, stroke, heart disease, emphysema, and other short- and long-term breathing problems. These troubling facts are very well-known to the public, both adults and adolescents. There is also a concerning association between the use of tobacco and other dangerous behaviors; e.g., the use of other substances such as alcohol and marijuana and high risk sexual behavior. Cognitive and mental problems including anxiety, depression and attention deficit/hyperactivity disorder are also associated with smoking. It is of great concern and frustration that all the above-mentioned facts do not prevent millions from starting to smoke every year around the world. In fact, smoking is practiced by every nation on the planet and appears to take place in every society, regardless of race, color or social status. In this book, we have gathered research from international collaborators to touch on these issues.

In: Public Health Concern   ISBN: 978-1-62948-424-2
Editors: J Merrick and A Tenenbaum   © 2014 Nova Science Publishers, Inc.

Chapter I

Smoking, alcohol and substance use

Ariel Tenenbaum, MD[*,1,2] and Joav Merrick, MD, Mmedsc, DMSc[1,2,3,4]

[1]Division of Pediatrics, Hadassah Hebrew University Medical Center,
Mt Scopus Campus, Jerusalem
[2]National Institute of Child Health and Human Development, Jerusalem
[3]Office of the Medical Director, Health Services,
Division for Intellectual and Developmental Disabilities,
Ministry of Social Affairs and Social Services, Jerusalem, Israel
[4]Kentucky Children's Hospital, University of Kentucky,
Lexington, Kentucky, US

Introduction

Smoking is the most common addiction among adolescents and young adults (1,2). In almost 90% of adults it is reported that the first use of tobacco was during adolescence. Although the use of tobacco has been declining over the last decades it remains one of the major health and social challenges and its use is still very common – among high school students almost one third of

[*] Correspondence: Ariel Tenenbaum, MD, Down Syndrome Center, Division of Pediatrics, Hadassah Hebrew University Medical Center, Mt Scopus Campus, POB 24035, IL-91240 Jerusalem, Israel. E-mail: tene@hadassah.org.il.

females and one half of males report using more than one tobacco product in the last 30 days. In the United States alone almost 400,000 young people become regular smokers every year (1,2).

Smoking has well known health hazards. It is associated with lung cancer, oral cancer, stroke, heart disease, emphysema, and other short and long term breathing problems. These troubling facts are very well known to the public, both adults and adolescents. There is also a concerning association between use of tobacco and other dangerous behaviors e.g. use of other substances such as alcohol and marijuana and high risk sexual behavior. Cognitive and mental problems including anxiety, depression and attention deficit/hyperactivity disorder are also associated with smoking (3). It is of great concern and frustration that all the above mentioned facts does not prevent millions from starting to smoke every year around the world. In fact smoking is practiced by every nation on the planet and appears to take place in every society, regardless of race, color or social status.

Some adolescents would pick up smoking because of false notions of presumably positive effects of smoking e.g. lower appetite resulting in weight loss and stress reduce effect. The ugly truth is that smoking is associated with undesirable cosmetic and social effects including yellow teeth, bad smell, bad breath and lower sport achievements.

There are social and environmental factors and trends that may increase smoking prevalence in adolescence. The history of tobacco use shows that males were always more likely to smoke compared with females (1,2). Advertisements encouraging people to smoke are still quite common around the world in newspapers, commercial breaks on TV and street posters, and are generally successful in increasing sales and use. The appearance of people smoking in movies and television shows may increase smoking rates, and family members and friends who smoke appear to have a similar effect. It is sad and even bizarre to witness family members, the young and the adults, smoking together even when obvious medical problems have already compromised the parents' and the grandparents' health and physical abilities (4). Smoking is more common among adolescents of low socioeconomic and educational status, while regular physical activity may be a preventive factor.

Smoking is addictive. Nicotine, a major ingredient in the tobacco, is highly addictive. Stopping this hazardous habit is therefore hard to do. Many products that contain nicotine and other substances were developed over the years and may help in quitting. However avoiding cigarettes is still much more effective.

Intervention to stop smoking

Intervention programs that aim to prevent adolescents from smoking or to quit smoking are introduced frequently, with considerable success (5). For example prohibition of tobacco products advertisements, preventing actors from smoking in films, reducing availability of these products in the market and raising their prices and taxes, and special educational programs in schools are some ways to intervene. Legal measures such as prevention of sale of tobacco products to minors are also used to reduce cigarette smoking.

In conclusion smoking is common among adolescents and its detrimental effects on the health and mental state of the smokers are well known. Smoking poses a major social, medical and economical challenge in every country in the world, and therefore adequate resources are required to fight against this hazardous behavior and thus preventing major diseases and disabilities.

References

[1]   US Department of Health and Human Services. Preventing tobacco use among youth and young adults: A report of the Surgeon General. Atlanta, GA: US Department Health Human Services, Centers Disease Control Prevention, Office Smoking Health, 2012.

[2]   Centers for Disease Control and Prevention. Current tobacco use among middle and high school students. United States, 2011. MMWR 2012;61(31):581-5.

[3]   Zoloto A, Nagoshi CT, Presson C, Chassin L. Attention deficit/hyperactivity disorder symptoms and depression symptoms as mediators in the intergenerational transmission of smoking. Drug Alcohol Depend 2012;126(1-2):147-55.

[4]   Gilman SE, Rende R, Boergers J, Abrams DB, Buka SL, Clark MA, et al. Parental smoking and adolescent smoking initiation: an intergenerational perspective on tobacco control. Pediatrics 2009;123(2):e274-81.

[5]   Centers for Disease Control and Prevention. Best practices for comprehensive tobacco control programs, 2007. Atlanta, GA: US Department Health Human Services, Centers Disease Control Prevention, National Center Chronic Disease Prevention Health Promotion, Office Smoking Health, 2007.

Section 1: Smoking

In: Public Health Concern
Editors: J Merrick and A Tenenbaum

ISBN: 978-1-62948-424-2
© 2014 Nova Science Publishers, Inc.

Chapter II

Smoking among adolescents living in the West Bank (Judea and Samaria)

Emmanuel Rudatsikira, Adamson S Muula[*]
and Seter Siziya
School of Community and Environmental Health,
Old Dominion University, Norfolk, Virginia, US,
Department of Community Health, University of Malawi,
College of Medicine, Blantyre, Malawi
and Department of Community Medicine, University of Zambia,
School of Medicine, Lusaka, Zambia

While our knowledge on the prevalence and associated factors of cigarette smoking continues to grow, there are limited data from the conflict regions of the Middle East. We therefore conducted a secondary analysis of the Global Youth Tobacco Survey (2005) from the West Bank (Judea and Samaria) to assess the prevalence and social correlates of cigarette smoking. Methods: Data from the Global Youth Tobacco Survey (GYTS) 2005 were used to determine current cigarette smoking prevalence and associated factors, attitudes, and exposure to, tobacco advertisements among adolescents. Current cigarette smoking was

---

[*] Correspondence: Adamson S Muula, Department of Community Health, University of Malawi, College of Medicine, Blantyre, Malawi. E-mail: amuula@medcol.mw.

defined as having smoked even a single puff within the last 30 days preceding the survey. Data were analyzed using SUDAAN software 9.0. Results: Of the 2,182 respondents, 28.7% males and 9.6% females reported being current smokers (p <0.01). Having smoking friends was strongly associated with smoking after controlling for age, gender, parental smoking status, and perception of risks of smoking (OR=9.79; 95% CI [6.75, 14.22] for most friends and OR=4.68; 95% CI [3.36, 6.52] for some friends). Male gender and having one or both smoking parents were associated with smoking (OR=2.29; 95% CI [1.70, 3.09] and OR=1.85; 95% CI [1.41, 2.42] respectively). Conclusions: We report that overall about 1 in 5 school-going adolescents in the West Bank were current cigarette smokers in 2005. Factors associated with smoking were having parents or peer who was smoker and feelings that smoking was harmful was protective. Many adolescents are exposed to pro-tobacco advertisements. In general, males who smoked were perceived positively than females.

Introduction

Tobacco use is the single most common cause of preventable morbidity and mortality from non-communicable disease (1-3). Many adults who smoke initiated the habit of tobacco use as adolescents (4,5). While many of the health effects of tobacco are experienced later in life, some health problems such as incident and exacerbation of asthma are observed in the short term (6,7).

Several authors have also reported on the clustering of unhealthy behaviours among adolescents who smoke [8-10]. These authors report that adolescent smokers are also likely to use illicit drugs, alcohol, and be truant from school. Thus cigarette smoking may be a marker of many other public health concerns that affect adolescents.

The West Bank (Judea and Samaria) is administratively located in the Eastern Mediterranean Region (EMR) of the World Health Organization (WHO). Warren et al (11) have reported that prevalence of current cigarette smoking among in-school adolescents in the Eastern Mediterranean region as 6.7% and 3.2% for males and females. In 2001, prevalence of current cigarette smoking among 13 to15 year old adolescents ranged from 13.9% to 14.7% in this region (12). While the Tobacco Youth Survey Collaborative Group (12) reported on the prevalence of tobacco use in West Bank, the report did not report on the correlates of cigarette smoking. The Tobacco Youth Survey Tobacco Group also only reported prevalence of tobacco use among 13 years to 15 years olds. Prevalence data, while useful may be limited as these on their own may fail to inform the design and evaluation of interventions aimed to prevent adolescent smoking. We therefore conducted secondary analysis of the West Bank GYTS-2005 to examine the prevalence and associated factors to cigarette smoking among in-school adolescents.

Current study

The current study involved secondary analysis of existing data available from the Centers for Disease Control and Prevention (CDC), Atlanta, Georgia, United States of America on West Bank Global Youth Tobacco Survey GYTS) conducted in 2005. Comprehensive description of the GYTS purpose and methodology has been reported elsewhere (13-16). In brief however, the GYTS is a school-based survey of students aged 13–15 years. It is a cross-sectional study utilizing a multistage sample design with schools selected proportional to enrollment size. Within a selected school, classrooms are chosen randomly. All the students within the selected classes are eligible for participation regardless of their actual ages. A questionnaire is self-completed anonymously by the students and this takes between 30 to 40 minutes. The GYTS core questionnaire aims to collect the following information or data to estimate the following: prevalence of cigarette smoking and other tobacco use among young people; knowledge and attitudes of young people towards cigarette smoking; role of the media and advertising on young people's use of cigarettes; access to cigarettes; tobacco-related school curriculum; exposure to environmental tobacco smoke (ETS) and cessation of cigarette smoking.

The West Bank GYTS used a two-stage probability sampling technique. In the first stage, primary sampling units were schools which were selected with a probability proportional to their enrolment size. In the second step of sampling, a systematic sample of classes in the selected school was obtained. All students in the selected classes were eligible to participate. A self-completed questionnaire was used and completed within one class session. Response rate in the West Bank GYTS was 96.5%. For the purpose of this study, only data related to estimation of prevalence of smoking and associated factors are reported.

Data analysis was performed using SUDAAN software version 9.0 (Research Triangle Institute, Research Triangle Park, NC, United States of America) (17) to calculate weighted prevalence estimates and 95% CIs. The main outcome was current cigarette smoking which was defined as having smoked cigarettes, even a single puff within the past 30 days of the preceding the date of completion of questionnaire. This is the standard definition that has been used within the Global Youth Tobacco Survey (13-16,18). Differences in estimates were judged as statistically significant at the $p<0.05$ level. We also conducted logistic regression analysis to estimate the association between relevant predictor variables and current smoking.

Findings

A total of 2,182 students participated in the study of whom 1002 (50.7%) were female, and 1094 (49.3%) were males. The median age was 15 years, range 11 to 17 years.

Table 1. Factors associated with current smoking among adolescents in Judea and Samaria

Characteristics	Percent of smokers	Unadjusted odds ratios with 95% CI	Adjusted odds ratios with 95% CI
Age (years)			
11-12	18.4	1.00	1.00
13	12.4	**0.63 [0.41, 0.96]**	0.93 [0.55, 1.57]
14	20.0	1.11 [0.77, 1.60]	1.34 [0.83, 2.16]
15	21.3	1.20 [0.84, 1.72]	1.50 [0.93, 2.42]
16-17	23.4	**1.43 [1.01, 2.03]**	1.53 [0.95, 2.45]
Gender			
Combined	18.9		
Female	9.6	1.00	1.00
Male	28.7	**3.81 [2.93, 4.96]**	**2.29 [1.70, 3.09]**
Parental smoking status			
None	31.3	1.00	1.00
One or both parents smokers	68.7	**2.01 [1.58, 2.55]**	**1.85 [1.41, 2.42]**
Best friend smokers			
None	6.8	1.00	1.00
Some	29.9	**5.89 [4.38, 7.92]**	**4.68 [3.36, 6.52]**
Most or all	49.6	**13.57 [9.71, 18.97]**	**9.79 [6.75, 14.22]**
Perception that smoking is harmful			
No	17.4	1.00	1.00
Yes	82.6	**0.66 [0.43, 1.02]**	0.79 [0.47, 1.32]

Table 1 presents the prevalence and factors associated with smoking. Of the 2,182 participants, 18.9 were current smokers (28.7% of males and 9.6% of females) reported being current smokers (p <0.001). Having smoking friends was strongly associated with smoking after controlling for age, gender, parental smoking status, and perception of hazards caused by smoking. Subjects who had smoking friends were more likely to smoke than those who had no smoking friends (OR=9.79; 95% [6.75, 14.22] for most friends smokers and OR=4.68; 95% CI [3.36, 6.52] for some smoking friends). Males were more than twice likely to smoke as females (OR=2.29; 95% [1.70, 3.09]). Compared to subjects with non smoking parents, those whose parents smoked were 1.85 more likely to smoke (OR=1.85; 95% CI [1.41, 2.42]).

Table 2 indicates the participants were exposed to tobacco adverts through billboards (77.8%), magazine (65.0%), and TV (49.4%). Almost one in five respondents (19.7%) reported having an item with cigarette brand logo on it. Compared to females, males had higher rates of those who reported having an item with cigarettes logo (p<0.001) and being exposed to tobacco adverts on billboards and magazine compared to females (p <001).

Table 2. Exposure to tobacco advertisements among adolescents in Judea and Samaria

Characteristics	Number of participants	% of total and 95% CI
		P=0.289
Seen cigarette brand name on TV in past 30 days	1571	49.4 [46.8, 51.9]
Males	864	51.4 [48.0, 54.9]
Females	707	47.2 [43.4, 50.9]
		P<0.001
Has item with cigarette brand logo	1985	19.7 [18.0, 21.6]
Males	1035	23.2 [20.6, 25.9]
Females	950	16.3 [14.0, 18.9]
		P<0.001
Seen tobacco adverts on billboards in past 30 days	2036	77.8 [69.7, 73.7]
Males	1056	76.7 [73.9, 79.2]
Females	980	67.1 [64.0, 70.0]
		P<0.001
Seen tobacco adverts in newspapers/magazines in past 30 days	2015	
Males	1049	71.1 [68.2, 73.8]
Females	966	59.2 [56.0, 62.3]

Table 3 indicates that the vast majority (92.1%) of the respondents felt that smoking is harmful. 53.3 % thought that male smokers had more friends while 33.4% thought so for females. 28.8% felt that boys who smoke are attractive and 22.7% felt the same for girls who smoke.

Table 3. Attitudes towards tobacco smoking distributed by gender among adolescents in Judea and Samaria

Characteristic	Number of participants	% of total and 95% CI
		P=0.231
Felt that boys who smoke have more friends	1459	53.3 [50.7, 55.9]
Males	823	54.5 [51.0, 58.0]
Females	636	51.8 [47.9, 55.8]
		P=0.130
Felt like girls who smoke had more friends	1512	33.4 [31.0, 35.9]
Males	824	31.4 [28.2, 34.8]
Females	688	35.5 [31.9, 39.2]
		P=0.842
Felt that boys who smoke are attractive	1531	28.9 [26.6, 31.3]
Males	809	28.7 [25.6, 32.0]
Females	722	29.1 [25.8, 32.6]
		P=0.487
Felt that girls who smoke are attractive	1645	22.7 [20.3, 24.8]
Males	873	23.1 [20.4, 26.2]
Females	772	22.2 [19.3, 25.3]
		P=0.079
Felt that tobacco smoking is harmful to health	1833	92.1 [90.7, 93.3]
Males	944	90.6 [88.4, 92.4]
Females	889	93.6 [91.7, 95.0]

Discussion

Our study of in-school adolescents in West Bank found an overall prevalence of current cigarette smoking at 18.9%. This is higher than prevalence estimates of 13.9% to 14.7% reported for 2001 among adolescents 13 to 15 years in the same area (12) . This difference could be due to a number of reasons. Firstly, in the 2001 study only study participants aged 13 to 15 years were reported. As age is associated with history of smoking, our inclusion of much older adolescents could partly explain this difference. The other possible reason is that the socio-political environment of the West Bank has changed over the years since 2001. The increased social disruptions experienced in the area may facilitate increasing smoking among adolescents. Also compared to the overall prevalence of adolescents' cigarette smoking in the Eastern Mediterranean region (6.7% and 3.2% for males and females), our finding suggest a higher prevalence of current cigarette smoking in the West Bank. The prevalence of smoking in this study is slightly higher than 16.6% reported from Jordan (12).

We found that males were likely to be smokers than females (OR=3.81, 95% CI, 2.93, 4.96). We believe this due to cultural acceptability of cigarette smoking amongst males than females. Rudatsikira et al (19) have reported male predominance of smoking among adolescents in Ethiopia. However, the Global Youth Tobacco Survey Collaborating Group has reported that overall, considering all the many settings that they have conducted the GYTS, there was little gender differences in cigarette smoking (20). This does not suggest that local differences do not occur; just the general picture does not support marked gender disparities.

We also found that having parents or friends who smoked was positively associated with the adolescents being a smoker. Believing that smoking was harmful to health was negatively associated with being a current smoker. The association between having smoking parents or friends and adolescents' own smoking has been demonstrated in many other studies (19,21-23). Mercken et al have reported that social selection and social influence play an important role in explaining similarity of smoking behaviour among friends (24). Mercken et al (24) have also reported that although parental smoking was important in influencing adolescent smoking, sibling smoking was much more influential. De vries et al (25) have reported also that adolescents' own attitudes towards smoking may influence them in the selection of friends i.e. adolescents with favourable attitudes towards smoking are more likely to choose smokers as friends.

We also assessed exposure of study participants to pro-tobacco advertisements through mass media. Overall as shown in Table 2, exposure to cigarette brand names on television, having an item with a tobacco brand logo, having seen billboard with tobacco advertisement was high. Males were more likely to have been exposed to these advertisements than females. As tobacco advertisements have been known to influence adolescents smoking (26-28), the higher prevalence of smoking among males compared to females could in part be explained by the higher exposure to pro-tobacco advertisements among the males.

While we found that there was a gender difference in exposure to pro-tobacco advertisements, there was no gender difference in attitudes towards smoking (Table 3). Over 90% of both males and females thought that smoking was harmful to health. Overall 53.3% thought boys who smoke had more friends, while 33.4% thought girls who smoked had more friends. 28.9% thought that boys who smoked were more attractive while 22.7% thought girls who smoked were more attractive. The fact that more study participants felt boys who smoked had more friends or were more attractive most likely exhibits societal attitudes that smoking is more acceptable among males than females.

Our study has a number of limitations. We used self-reported information on current smoking status. Data on smoking status within the GYTS methodology are not validated with a biomarker such as exhaled carbon monoxide or cotinine levels (29-30). The degree of potential reporting bias was not assessed. However data from the United States using similar methodology have shown good test-retest reliability (31). Huerta et al have also reported high reliability of smoking history among Israeli youths (32). However these authors noted that as years pass from date since smoked, data become less reliable. Our study is based on smoking history within the last 30 days and so should be reasonably reliable.

The data used obtained from the survey were collected from school-going adolescents. To the extent that the smoking prevalence and associated factors between in-school and out-of school adolescents, our finding may not be generalized to the whole adolescent population in West Bank.

Data were also collected from adolescents who were present at school on the day of the survey. These absent students may have different characteristics from those who are present. Michaud et al (32) have reported in a survey of adolescents in Switzerland that higher proportion of absent students were sexually active, used tobacco, alcohol, and cannabis more often than did present students. Thus to the extent that absent students are different from present peers in West Bank, bias may have been introduced. However the with a response rate of 96.5%, any bias that may have resulted because of absenteeism is likely to have been minimal.

Conclusion

We have reported that overall about 1 in 3 school-going adolescents in the West Bank were current cigarette smokers in 2005. Factors associated with smoking were having parents or peer who was smoker and feelings that smoking was harmful was protective. Many adolescents in the West Bank are exposed to pro-tobacco advertisements. In general, males who smoked were perceived positively than females.

Acknowledgments

The authors thank the Centers for Disease Control and Prevention (CDC), Atlanta, Georgia, United States of America, for making the data used in this study available for our analysis. However, the CDC took no part in the decision of the models used and any other data analysis or the decision to publish this paper. We also thank the students who participated in the survey. The authors declare they have no competing interests. Authors' contribution: ER designed analysis plan, led data analysis and participated in drafting of manuscript, SS participated in the interpretation of findings and drafting of manuscript and ASM participated in the interpretation of findings and drafting of manuscript.

References

[1]   Ezzati M, Lopez AD, Rodgers A, Vander Hoorn S, Murray CJL and the Comparative Risk Assessment Collaborating Group. Selected major risk factors and global and regional burden of disease. Lancet 2002; 360:1347-60.

[2]   Peto R, Lopez AD, Boreham J, Thun, Heath Jr C. Mortality from smoking in developed countries 1950–2000: indirect estimation from National Vital Statistics. Oxford: Oxford University Press, 1994.

[3]   WHO IARC. Monographs on the evaluation of carcinogenic risks to humans, vol 83: tobacco smoke and involuntary smoking. Lyon: IARC Press, 2004.

[4]   Brook JS, Balka EB, Ning Y, Brook DW. Trajectories of cigarette smoking among African Americans and Puerto Ricans from adolescence to young adulthood: associations with dependence on alcohol and illegal drugs. Am J Addict 2007;16:195-201.

[5]   White HR, Violette NM, Metzger L, Stouthamer-Loeber M. Adolescent risk factors for late-onset smoking among African American young men. Nicotine Tob Res 2007;9:153-61.

[6]   Gilliland FD, Islam T, Berhane K, Gauderman WJ, McConnell R, Avol E, Peters JM. Regular smoking and asthma incidence in adolescents. Am J Respir Crit Care Med 2006;174:1094-100.

[7]   Annesi-Maesano I, Oryszczyn MP, Raherison C, Kopferschmitt C, Pauli G, Taytard A, et al. Increased prevalence of asthma and allied diseases among active adolescent tobacco smokers after controlling for passive smoking exposure. A cause for concern? Clin Exp Allergy 2004; 34:1017-23.

[8]   Dierker LC, Sledjeski EM, Botello-Harbaum M, Ramirez RR, Chavez LM, Canino G. Association between psychiatric disorders and smoking stages within a representative clinic sample of Puerto Rican adolescents. Compr Psychiatry 2007; 48:237-44.

[9]   Rhee D, Yun SC, Khang YH. Co-occurrence of problem behaviors in South Korean adolescents: findings from Korea Youth Panel Survey. J Adolesc Health 2007;40:195-7.

[10]   Ohene SA, Ireland M, Blum RW. The clustering of risk behaviors among Caribbean youth. Matern Child Health J 2005;9:91-100.

[11]   Warren CW, Jones NR, Eriksen MP, Asma S. Global Tobacco Surveillance System (GTSS) collaborative group: Patterns of global tobacco use in young people and implications for future chronic disease burden in adults. Lancet 2006;367:749-53.

[12]   The Global Youth Tobacco Survey Collaborative Group: Tobacco use among youth: a cross country comparison. Tobacco Control 2002;11: 252-270.

[13]   Warren CW, Riley L, Asma S, Eriksen MP, Green L, Blanton C, et al. Tobacco use by youth: a surveillance report from the Global Youth Tobacco survey project. Bull World Health Organ 2000;78(7):868–76.

[14]   The Global Youth Tobacco Survey Collaborative Group. Tobacco use among youth: a cross country comparison, Tob Control 2002;11:252-70.

[15]   The Global Youth Tobacco Survey Collaborative Group. Differences in worldwide tobacco use by gender: findings from the Global Youth Tobacco Survey. J Sch Health 2003;73:207–215.

[16]   Global Tobacco Surveillance System Collaborating Group. Global Tobacco Surveillance system (GTSS): purpose, production, and potential. J Sch Health 2005;75:15–24.

[17]   Shah BV, Barnwell BG, Bieler GS. SUDAAN: software for the statistical analysis of correlated data. User's manual, release 7.5. Research Triangle Institute, Research Triangle Park, 1997.

[18]   Kyrlesi A, Soteriades ES, Warren CW, Kremastinou J, Papastergiou P, Jones NR, et al. Tobacco use among students aged 13-15 years in Greece: the GYTS project. BMC Public Health 2007;7:3.

[19]   Rudatsikira E, Abdo A, Muula AS. Prevalence and determinants of adolescent tobacco smoking in Addis Ababa, Ethiopia. BMC Public Health 2007;7:176.

[20]   Global Youth Tobacco Survey Collaborating Group. Differences in worldwide tobacco use by gender: findings from the Global Youth Tobacco Survey. J Sch Health 2003;73:207-15.

[21]   Jindal SK, Aggarwal AN, Gupta D, Kashyap S, Chaudhary D. Prevalence of tobacco use among school going youth in North Indian States. Indian J Chest Dis Allied Sci 2005;47:161-6.

[22] Scragg R, Laugesen M. Influence of smoking by family and best friend on adolescent tobacco smoking: results from the 2002 New Zealand national survey of year 10 students. Aust NZ J Public Health 2007; 31:217-23.

[23] Livaudais JC, Napoles-Springer A, Stewart S, Kaplan CP. Understanding Latino adolescent risk behaviors: parental and peer influences. Ethn Dis 2007;17:298-304.

[24] Mercken L, Candel M, Willems P, de Vries H. Disentangling social selection and social influence effects on adolescent smoking: the importance of reciprocity in friendships. Addiction 2007;102:1483-93.

[25] de Vries H, Candel M, Engels R, Mercken L. Challenges to the peer influence paradigm: results for 12-13 year olds from six European countries from the European Smoking Prevention Framework Approach study. Tob Control 2006;15:83-9.

[26] Andreeva TI, Krasovsky KS, Semenova DS. Correlates of smoking initiation among young adults in Ukraine: a cross-sectional study. BMC Public Health 2007;7:106.

[27] Peters RJ, Kelder SH, Prokhorov A, Springer AE, Yacoubian GS, Agurcia CA, Amos C: The relationship between perceived exposure to promotional smoking messages and smoking status among high school students. Am J Addict 2006;15:387-91.

[28] DiFranza JR, Wellman RJ, Sargent JD, Weitzman M, Hipple BJ, Winickoff JP. Tobacco Consortium, Center for Child Health Research of the American Academy of Pediatrics: Tobacco promotion and the initiation of tobacco use: assessing the evidence for causality. Pediatrics 2006;117:e1237-48.

[29] Thaqi A, Franke K, Merkel G, Wichmann HE, Heinrich J. Biomarkers of exposure to passive smoking of children: frequency and determinants. Indoor Air 2005;15:302-10.

[30] Breland AB, Kleykamp BA, Eissenberg BA. Clinical laboratory evaluation of potential reduced exposure products for smokers. Nicotine Tob Res 2006;8:727-38.

[31] Brener ND, Colins JL, Kann L, Warren CW, Williams BT. Reliability of the Youth Risk Behavior Survey Questionnaire. Am J Epidemiol 1995; 141:575-80.

[32] Huerta M, Chodick G, Balicer RD, Davidovitch N, Grotto I. Reliability of self-reported smoking history and age at initial tobacco use. Prev Med 2005;41:646-50.

[33] Michaud PA, Delbos-Piot I, Narring F. Silent dropouts in health surveys: are nonrespondent absent teenagers different from those who participate in school-based health surveys? J Adolesc Health 1998;22:326-33.

In: Public Health Concern   ISBN: 978-1-62948-424-2
Editors: J Merrick and A Tenenbaum   © 2014 Nova Science Publishers, Inc.

Chapter III

Smokeless tobacco use among university students

Mohammad Ali Emamhadi, MD[*1],
Maryam Jalilvand, MSc[2], Zahra Nikmanesh, PhD[3]
and Yahya Kazemi, PhD[3]

[1]Clinical Toxicology and Forensic Medicine, School of Medicine,
Shahid Beheshti University of Medical Sciences, Tehran
[2]Training and Education Office of Tehran Province, Tehran
[3]Sistan-Balouchestan University, Zahedan, Iran

It is well known that tobacco is a cause of morbidity and mortality. Use of smokeless tobacco such as Paan, Nass, Gutka and Tumbaku is common in South and Southeast Asia, but not in Iran and here it is not considered a normal cultural habit. A cross sectional study was carried out in five colleges of Sistan Baloochestan, the Southeastern province of Iran. Three hundred fifty four students selected by convenience sampling completed a peer reviewed, pre-tested, self-administered questionnaire. Thirty nine (11.0%) students were lifetime users of smokeless tobacco among which nineteen (5.4%) occasional users, seven (2.0 %) current users and thirteen (3.6%) fulfilled the criterion for established users. Paan was the most commonly used form of smokeless tobacco followed by

* Correspondence: Associate professor Mohammad Ali Emamhadi, MD, Clinical Toxicology and Forensic Medicine, School of Medicine, Shahid Beheshti University of Medical Sciences, POBox 13185-1678, Tehran, Iran. E-mail: swt_f@yahoo.com

Nass. On univariate analysis, lifetime use of smokeless tobacco showed significant associations with the use of cigarettes, student gender (M > F), individual condition (native > guest) and kind of the college (engineering > psychology). The use of smokeless tobacco among students should not be ignored. The governments should add preventive measures against smokeless tobacco use to the existing law against cigarette smoking.

Introduction

Tobacco can be used in ways ranging from cigarette and cigar to chewing of 'smokeless tobacco'. This latter category includes various forms of tobacco with Paan/betel quid being the most commonly used (1). Most people place these in the mandibular or labial groove and suck on them slowly for 10–15 minutes or simply apply them to their teeth and gums (the same). The culturally popular product is "Paan", which consists of a number of ingredients, including tobacco, areca nut, slaked lime, and spices. Nass is a commercially powdered mixture containing the same ingredients as Paan. It was introduced in India nearly three decades ago. Paan, Nass and gutka are widely used in south and Southeast Asia and by those immigrants, and their use has spread across to other countries (2-12). It is said that betel quid and chewable tobacco is the fourth most commonly used psycho-active substance in the world, ranking after caffeine, alcohol and nicotine (1). Having an ancient history, they are an integral part of the culture and sometimes erroneously believed to have medicinal benefits (1,13).

Smokeless tobacco users in India and Pakistan together have been estimated to number 100 million (14). In India about 35–40% of tobacco consumption is in smokeless forms, while an earlier study in Pakistan showed that 21% of men and 12% of women were users of betel quid (15). In Pakistan, a recent study among the adolescents and adults of a Karachi squatter settlement reported that 40% of the population was using at least one chewable product of betel, areca and tobacco on a daily basis (16).

Unfortunately it is reported that increasing use occurred among vulnerable groups such as children, teenagers, women, immigrants of South Asian to other countries and also adjacent countries especially boundary's states (1).

Paan contains areca nut, betel leaf and calcium hydroxide but tobacco and various other spices are also commonly added (17). Industrially prepared mixture of areca nut, lime, catechin containing substance, sandalwood fragrance with tobacco (gutka) or without tobacco (chaalia) were introduced in recent decades, which have contributed to growth and use of these products (1). Tumbaku and naswar mainly contain tobacco with small amount of spices, areca and betel. Tumbaku is oral chewable form of tobacco while naswar is placed in oral vestibule (18).

Factors that continue to encourage people to use smokeless tobacco include its affordability, ease of purchase or production and the widely held misconception that it has medicinal value for improvement in tooth ache, headache and stomach ache (19). Furthermore, in contrast to cigarettes, there is no taboo against using smokeless tobacco and the government efforts have also focused more on eliminating cigarette use than tobacco as a whole (13,19). All these, coupled with peer pressure and the belief that smokeless tobacco is less hazardous than cigarette smoking mean that these forms continue to be used by vast numbers of people.

Presently tobacco use is the leading preventable cause of death globally (20) and it is estimated that by 2030, it would account for over 10 million annual deaths worldwide (21,22)), 70% of which will be in the developing world (23). All forms of tobacco carry serious health consequences, most importantly oral and pharyngeal cancers (24-26) and other malignancies of the upper digestive tract (22,27). Other ingredients combine with tobacco to produce a product with an even higher carcinogenicity for humans.

Chewing betel, areca and smokeless tobacco products lead to discoloration of teeth, development of chronic debilitating diseases involving gingival and oral mucosa, and higher mortality among users. These diseases include oral sub mucous fibrosis, oral leukoplasia, oral cavity and other head and neck cancers (1,28,29).

There is also evidence that smokeless tobacco is a risk factor for hypertension and dyslipidemia (19). Chewing of tobacco by pregnant mothers has been found to cause an increased incidence of still births and low birth weight deliveries (the same). In addition, chewing of betel quid, with or without tobacco can aggravate asthma and predispose the users to diabetes mellitus (the same). Regular use of Paan and gutka leads to oral cancer and precancerous conditions (30-34).

It seems the growing popularity of Paan use in eastern province of Iran has troubled the public health community. The extent of Paan and Nass use in Iran and the resultant consequences have not yet been studied. Iran is a neighbor of Pakistan where smokeless tobacco use is legal and easily available. Until recent years, smokeless tobacco was very rare in Iran, but now its use is spreading slowly in these parts of the country. It is shown that costs and consequences of tobacco use impose a heavy social and economic burden on a nation. Much of this can be avoided by policies and awareness programs aimed at reducing tobacco use. With this background, this chapte was conducted with the objective of determining the prevalence of smokeless tobacco use among university students from five different colleges in Sistan-Balouchestan province (in southeastern Iran, adjacent to Pakistan) as a first step to address this gap in information about the exact extent of Paan or smokeless tobacco use in Iran, especially in the eastern provinces.

The purpose of this chapter is to explore the patterns of Paan and Nass consumption in the eastern states such as Sistan-Balouchestan. These types of studies have important implications for future smokeless tobacco investigations and interventions in these large, growing communities.

Our study

This was a multi-center cross sectional study carried out on students of five colleges of Sistan-Balouchestan University during the period April-July 2008. The colleges (engineering, psychology, management, science and literature) were selected in order to compare the patterns of tobacco use in students with different conditions.

We chose all students of those colleges, nearly 381 cases out of which 354 subjects filled the questionnaire, while 27 refused (response rate = 92.9%). Convenience sampling was used for getting the questionnaires filled during regular college hours by students.

A peer reviewed, pilot tested, anonymous self-administered questionnaire was used. Questions were asked regarding occasional, current, and established use of smokeless tobacco. Occasional users were defined as having used smokeless tobacco at least once or twice in their life. Current users were defined as having used smokeless tobacco at least once in the last 30 days while established users were defined as having used smokeless tobacco on more than 100 occasions in their lifetime. Questions were also asked regarding the form of smokeless tobacco they used, any cigarette smoking, as well as the age at which they took up these habits. Relevant demographic information was also obtained.

Ethical approval for the study was obtained from the Ethical Committee of Sistan-Balouchestan University. The nameless questionnaires were collected back in an unmarked envelope to ensure complete confidentiality. The study was conducted in compliance with the 'Ethical Principles for Medical Research involving Human Subjects' of Helsinki Declaration (35). Verbal informed consent was obtained from all subjects and documented in the presence of a witness.

Data was entered and analyzed with Statistical Package for Social Sciences (SPSS) version 15. Descriptive statistics of socio-demographic information and use of chewable tobacco products were obtained. Univariate and multivariate odds ratio with 95 percent confidence interval were obtained using Chi square and logistic regression analyses, respectively. For all purposes, a p value of < 0.05 was considered to be significant.

Findings

Of the 354 students that completed the questionnaire, 170 (48.0%) were males and 184 (52.0%) were females. The mean age of the sample was 21.57 and its median was 21 years (SD: 1.90).

One hundred and forty six (41.24 %) students had used tobacco in some form (smoked or smokeless) in their lifetime. Thirty nine (11.01 %) students were lifetime users of smokeless tobacco of which seven (1.97 %) were current users while 13 (3.67 %) fulfilled the criterion for established users and nineteen (5.36 %) students were occasional users of smokeless tobacco .The frequency and form of smokeless tobacco use is shown in table 1.

Table 1. Pattern of use of smokeless tobacco among colleges' students

		Frequency	Percent
	Occasional	19	5.36
Situation	Daily	7	1.98
	Established	13	3.67
Kind	Paan	27	7.63
	Nass	12	3.38
Non Smokeless user		315	88.99
Total		354	100.00

Paan (7.63 %) was the most commonly used form of smokeless tobacco followed by Nass (3.38 %). About 21 (%78) individuals of Paan users belonged to the Engineering College while 7 (%52) of Nass users studied in Psychology College.

About 30 (%77) of lifetime users also smoked cigarettes while among people who had never used smokeless tobacco, only 12(%30.5) were smokers (p value: < 0.001). The mean age at which the students began smoking, was 17.94 (SD=1.64) years while the mean age at which they began using smokeless tobacco was 18.14 (SD=1.49) years, which was not significant; p value: 0.29.

Lifetime use of smokeless tobacco was also found to have significant associations with student gender (M > F, p value: < 0.000), student habitat (native > guest, p value: 0.07) and kind of College (Engineering > Psychology, p: < 0.001). The frequency of demographic data in smokeless tobacco lifetime users is shown in table 2.

Table 2. Demographic data in lifetime user of smokeless tobacco

		Frequency	Percent
Gender	Male	36	10.17
	Female	3	0.84
	Engineering	16	4.52
	Psychology	11	3.11
College	Management	3	0.84
	Science	5	1.41
	Literature	4	1.13
Habitat	Native	29	8.19
	Guest	10	2.82

Multivariate analysis showed that there was a higher prevalence of smoking among students who were lifetime users versus those who had not used smokeless tobacco (O.R: 4.203 [2.279–7.751], p value: < 0.000). This association was independent of age, gender, habitat (being native) and kind of college.

Gender was also found to be independently associated with lifetime use of smokeless tobacco. Male students were more likely to be lifetime users than female students. (O.R: 2.198 [1.177–4.102], p value: 0.002).

An independent association was also found between lifetime use of smokeless tobacco and the kind of college. There was a higher prevalence of lifetime users among students from the Engineering College compared to Psychology. (O.R: 2.155 [1.250–3.716], p value: < 0.008). Results of multivariate analysis are shown in table 3.

**Table 3. Predictors of lifetime use of smokeless tobacco
on multivariate analysis**

	O.R.	C.I.	p-value
Gender	2.198	1.77 – 4.102	0.002
Location of College	2.155	1.250 – 3.716	0.008
Cigarette smoking	4.203	2.279 – 7.751	0.000

Discussion

To the best of our knowledge, this pilot study was the first evaluation of Paan and Nass use in Iran. Almost all studies carried out in Iran, have focused on the patterns of cigarette smoking alone and not the use of smokeless tobacco. Although some studies have been carried out, especially in India and Pakistan about smokeless tobacco, because of its common use in those countries, regarding its role in head and neck cancers especially oral cavity and pharynx (24,26,28,36), bladder carcinoma (37), peptic ulcer disease (38) and oral mucosal lesion (26). In the studies carried out in India and Pakistan, the range of lifetime smokeless tobacco users was in the range of 23 to 47 percent (36-40). Understandably, our figure of 11 percent prevalence is much lower compared to the rates among patients with conditions likely to be the result of long term use of smokeless tobacco.

It is reported that in 1982, 21 percent of people in Karachi used Paan, and in a recent study (36) 40 percent used smokeless tobacco. Various studies (40-42) have shown that the use of smokeless tobacco is inversely associated with the level of education and this might explain the lower prevalence reported by our study because our population was comprised of guest students who were also likely to be more aware of the hazards of smokeless tobacco than others, but it should be mentioned that Sistan-Balouchestan is a boundary state.

Most studies have reported that Paan is the most common form of smokeless tobacco use in India, but in Pakistan Nass or Naswar were the more popular choices. Our study report Paan to be the most commonly used among students. More significantly, it was seen that 52 percent of Paan users belonged to the Engineering College. In Sistan-Balouchestan, despite neighboring Pakistan, Paan is a new drug and it is more popular than Nass, especially among educated people and guests. It is well known that most guest students are in Engineering College and the highest mean score of examination were observed in that college. On the other hand, industrially prepared Paan marketed in bright, attractive sachets with appealing brand names like 'Sir', 'Shahi (royal)', are gaining popularity, especially among the guest students.

We also reported an independent association between the use of smokeless tobacco and the kind of college, with students from Engineering College being more likely to be lifetime users. One explanation of this is the high tendency of those students toward courageous and risky behaviors.

Our study also showed a significantly higher prevalence of smoking among users of smokeless tobacco. This could be because the same risk factors probably encourage people to take up smoking as well as the use of smokeless tobacco. In our study, the mean age at which the students started smoking was similar to that at which the students began using smokeless tobacco. This means that both habits are acquired at an equal age, again signifying possible similar reasons behind the use of smoked and smokeless tobacco.

Our finding that the use of smokeless tobacco was more common among the male gender is in line with what was found by Imam et al (5) and Mazahir et al (6). We feel this is because the use of tobacco (smoke and or smokeless) remains socially more acceptable for males than females.

On univariate analysis, we found an association between native students and using smokeless tobacco and multivariate analysis showed that this was an independent association. In the other hand, common use of smokeless tobacco in native people in boundary states widely spreads among educated classes.

Use of smokeless tobacco by colleges' students, should not be ignored, considering their future role in communities. Adding the goal of eliminating the use of smokeless tobacco to the existing law against cigarette smoking may help. This is because similar factors seem to be promoting the use of cigarettes as well as smokeless tobacco. Also, colleges should consider providing greater education about the myths and hazards of smokeless tobacco. Furthermore, boundary states preferences for the smokeless tobacco use should be kept in mind while planning preventive programs. Law in boundary states has to focus more on eliminating Paan and Nass usage. Further community-based studies are required to highlight the health burden due to smokeless tobacco and to better plan anti-tobacco law in the existing resources of a developing country such as Iran.

This study is not representative of Iranian students' drug abuse situation or condition. We tried to focus our study just on boundary states students from five colleges of Sistan-Balouchestan (Southeast state, neighbor of Pakistan) for estimating the spread of smokeless tobacco usage.

Acknowledgment

The authors would like to thank Farzan Institute for Research and Technology for technical assistance.

References

[1]   Gupta PC, Ray CS. Epidemiology of Betel quid usage. Ann Acad Med Singapore 2004;33(Suppl 4):31S–6S.

[2]   Atwal GS, Warnakulasuriya KAAS, Gelbier S. Betel quid (pan) chewing habits among a sample of south Asians. J Dent Res 1996; 75:1151.

[3]   Bedi R, Gilthorpe MS. The prevalence of betel-quid and tobacco chewing among the Bangladeshi community resident in a United Kingdom area of multiple deprivation. Prim Dent Care 1995;2:39–42.

[4]   Changrani J, Gany FM, Cruz G, Kerr R, Katz R. Paan and Gutka use in the United States: A pilot study in Bangladeshi and Indian-Gujarati immigrants in New York City. J Immigrant Refuge Stud 2006;4:99–110.

[5]   Imam SZ, Nawaz H, Sepah YJ, Pabaney AH, Ilyas M, Ghaffar S. Use of smokeless tobacco among groups of Pakistani medical students - a cross sectional study. BMC Public Health 2007;7:231.

[6]   Mazahir S, Malik R, Maqsood M, Merchant KA, Malik F, Majeed A, et al. Socio-demographic correlates of betel, areca and smokeless tobacco use as a high risk behavior for head and neck cancers in a squatter settlement of Karachi, Pakistan. Substance Abuse Treat Prev Policy 2006;1:10.

[7]   Shetty KV, Johnson NW. Knowledge, attitudes and beliefs of adult south Asians living in London regarding risk factors and signs for oral cancer. Comm Dent Health 1999;16:227–31.

[8]   Seedat HA, van Wyk CW. Betel-nut chewing and submueous fibrosis in Durban. S Afr Med J 1988;74:568–71.

[9]   Vora AR, Yeoman CM, Hayter JP. Alcohol, tobacco and paan use and understanding of oral cancer risk among Asian males in Leicester. Br Dent J 2000;188:444–51.

[10]   Summers RM, Williams SA , Curzon MEJ. The use of tobacco and betel quid ('pan') among Bangladeshi women in West Yorkshire. Comm Dent Health 1994;11:12–6.

[11]   Warnakulasuriya KAAS, Trivedy C, Maher R, Johnson NW. Aetiology of oral submucous fibrosis. Oral Dis 1997;3:286–7.

[12]   Warnakulasuriya KAAS, Johnson NW. Epidemiology and risk factors for oral cancer: Rising trends in Europe and possible effects of migration. Int Dent J 1996;46:245–50.

[13]   Nishtar S, Ahmed A, Bhurgri Y, Mohamud KB, Zoka N, Sultan F, et al. Prevention and control of cancers: National Action Plan for NCD Prevention, Control and Health Promotion in Pakistan. J Pak Med Assoc 2004;54:S45–56.

[14] Croucher R, Choudhury SR. Tobacco control policy initiatives and UK resident Bangladeshi male smokers: community-based, qualitative study. Ethnic Health 2007;2:321-37.

[15] Mahmood Z. Smoking and chewing habits of people of Karachi–1981. J Pak Med Assoc 1982;32:34–7.

[16] Khawaja MR, Mazahir S, Majeed A, Malik F, Merchant KA, Maqsood M, et al. Knowledge, attitude and practices of a Karachi slum population regarding the role of products of Betel, Areca and smokeless tobacco in the etiology of head and neck cancers. J Pak Med Assoc 2005;Suppl:S41.

[17] Mack TM. The pan-Asian Paan problem. Lancet 2001;357:1638–9

[18] Johnson N. Tobacco use and oral cancer: A global perspective. J Dent Educ 2001;65:328–39.

[19] Gupta PC, Ray CS. Smokeless tobacco and health in India and South Asia. Respirology 2003;8:419–31.

[20] Brundtland GH. Achieving worldwide tobacco control. JAMA 2000;284:750–1.

[21] John RM. Tobacco consumption patterns and its health implications in India. Health Policy 2005;71:213–22.

[22] Warnakulasuriya KAAS, Sutherland G, Scully C. Tobacco, oral cancer, and treatment of dependence. Oral Oncol 2005;41:244–60.

[23] World Health Organization. Tobacco Free Initiative. Addressing the worldwide tobacco epidemic through effective evidence-based treatment. Expert Meeting, Rochester, MN, USA, 1999.

[24] Avon SL. Oral mucosal lesions associated with use of Quid. J Can Dent Assoc 2004;70:244–8.

[25] IARC. Monographs on the evaluation of the carcinogenic risk of chemicals to humans, Tobacco habits other than smoking: betel-quid and areca-nut chewing; and some related nitrosamines. Vol. 37. Intl Agency Res Cancer, Lyon: 1985:141–200.

[26] Merchant A, Husain SS, Hosain M. Paan without tobacco: an independent risk factor for oral cancer. Int J Cancer 2000;86:128–31.

[27] Bhurgri Y, Faridi N, Kazi LA, Ali SK, Bhurgri H, Usman A, et al. Cancer esophagus Karachi 1995–2002: epidemiology, risk factors and trends. J Pak Med Assoc 2004;54:345–8.

[28] Bhurgri Y, Bhurgri A, Hussainy AS, Usman A, Faridi N, Malik J, et al. Cancer of the oral cavity and pharynx in Karachi. Identification of potential risk factors. Asian Pac J Cancer Prev 2003;4:125–30.

[29] Tsai JF, Chuang LY, Jeng JE, Ho M S, Hsieh M Y, Lin ZY, et al. Betel quid chewing as a risk factor for hepatocellular carcinoma: a case-control study. Br J Cancer 2000;84:709–12.

[30] Babu S, Bhat RV, Kuma PU, Sesikaran B, Rao KV, Aruna P, et al. A comparative clinico-pathological study of oral submucous fibrosis in habitual chewers of pan masala and betelquid. J Toxicol Clin Toxicol 1996;34:317–22

[31] Johnson NW, Ranasinghe AW, Warnakulasuriya KAAS. Potentially malignant lesions and conditions of the mouth and oropharynx: Natural history-cellular and molecular markers of risk. Eur J Cancer Prev 1993:2:31–51.

[32] Mehta FS, Maner J. Leukoplakia in tobacco related oral mucosal lesions and conditions in India. Bombay: Basic Dental Res Unit, 1993:23–46.

[33] Murti PR, Bhonsle RB, Gupta PC. Etiology of oral submucous fibrosis with special reference to the role of areca nut chewing. J Oral Pathol Med 1995;24:145–52.

[34] Trivedy CR, Craig G, Warnakulasuriya S. The oral health consequences of chewing areca nut. Addict Biol 2002;7:115–25.

[35] The World Medical Association. World Medical Association Declaration of Helsinki: Ethical principles of medical research involving human subjects. Accessed 2010 Jul 14. URL: http://www.wma.net/en/30publications/10policies/b3/17c.pdf.

[36] Khawaja MR, Mazahir S, Majeed A, Malik F, Merchant KA, Maqsood M, et al. Chewing of betel, areca and tobacco: perceptions and knowledge regarding their role in head and neck cancers in an urban squatter settlement in Pakistan. Asia Pac J Cancer Prev 2006; 7:95-100.

[37] Rafique M. Clinico-pathological features of bladder carcinoma in women in Pakistan and smokeless tobacco as a possible risk factor. World J Surg Oncol 2005;3:53.

[38] Afridi MA. Tobacco use as contributory factor in peptic ulcer disease. J Collage Phys Surg Pakistan 2003;13:385–7.

[39] Jayakody AA, Viner RM, Haines MM, Bhui KS, Head JA, Taylor SJ, et al. Illicit and traditional drug use among ethnic minority adolescents in East London. Public Health 2006;120:329-38.

[40] Nisar N, Billoo N, Gadit AA. Pattern of tobacco consumption among adult women of low socioeconomic community Karachi, Pakistan. J Pak Med Assoc 2005;55:111–4.

[41] Gupta PC. Survey of sociodemographic characteristics of tobacco use among 99,598 individuals in Bombay, India using handheld computers. Tobacco Control 1996;5:114–20.

[42] Rani M, Bonu S, Jha P, Nguyen SN, Jamjoum, L. Tobacco use in India: prevalence and predictors of smoking and chewing in a national cross sectional household survey. Tobacco Control 2003; 12:4.

In: Public Health Concern   ISBN: 978-1-62948-424-2
Editors: J Merrick and A Tenenbaum   © 2014 Nova Science Publishers, Inc.

Chapter IV

Alcohol use, cigarette use, delinquency and Thai parenting practices

Brenda A Miller, PhD[*1]*, Hilary F Byrnes, PhD*[1],
Pamela K Cupp, PhD[2]*, Aphichat Chamratrithirong, PhD*[3],
Orratai Rhucharoenpornpanich, PhD[3],
Warunee Fongkaew, PhD[4]*, Michael J Rosati, MA, MEd*[5],
Warunee Chookhare, MBA[6]
and Rick S Zimmerman, PhD[7]

[1]Prevention Research Center, Pacific Institute for Research and Evaluation,
Berkeley, California, US
[2]Pacific Institute for Research and Evaluation, Louisville,
Kentucky, US
[3]Institute for Population and Social Research, Mahidol University,
Thailand
[4]Chiang Mai University, Chiang Mai, Thailand
[5]Thailand Ministry of Public Health, Department of Mental Health,
Rajanukul Institute, Bangkok, Thailand
[6]CSN and Associates, Bangkok, Thailand
[7]Virginia Commonwealth University, Richmond, Virginia, US

---

[*] Correspondence: Brenda A Miller, PhD, Prevention Research Center, 1995 University Ave, Suite 450, Berkeley, CA 94704, United States. E-mail: bmiller@prev.org.

For this chapter data were obtained from face-to-face interviews conducted with 420 randomly selected families (one parent, one 13-14 year old teen) in their homes from seven districts of Bangkok, Thailand. Adolescent risky behaviors that may be influenced by parenting practices and family rituals include alcohol use, cigarette use, and delinquency. Measures include: parental monitoring, parenting style, parental closeness, parental communication, and family rituals. Findings reveal increased alcohol use among Thai adolescents exposed to risks in family rituals. Lower prevalence of cigarette use is indicated among youth exposed to authoritative parenting and greater levels of parental monitoring. Serious delinquency is related to more risks in family rituals, but for girls only. Minor delinquency is related to less rule-setting, but also for girls only. These analyses provide support for using a risk and protective framework for guiding prevention strategies in Thailand. The relationship between family rituals and adolescent behaviors warrants further investigation and especially the elements of family rituals that reflect positive vs. the negative forces in the family dynamics.

Introduction

In Thailand families are migrating out of rural areas to cities and the stresses of urban life, influences of international tourists, and less availability of extended families create conditions that may negatively impact the ability to parent (1). In contrast, many western countries have experienced the transition to an industrialized state over more years, allowing parents and families to adjust more slowly. Parenting and family practices that evolve gradually over time, may allow better responses to the demands created by the historical and developmental events within the country. However, when changes occur more rapidly, families may have more difficulty in adjusting.

Understanding parent practices in Thailand requires understanding cultural influences that can provide support and protection to youth. For example, religious beliefs and traditions are important components of Thai family life that may be impacted (1). Families are less integrated in their neighborhoods as urban lifestyles are more varied and less closely knit than rural communities. Urban life is more influenced by western influences. Family dynamics that have characterized the Thai culture have been impacted by these urban strains. Traditionally, hierarchical relationships characterized families but this is changing (1). There exists a tension between traditional cultural values and the changes evolving in modern society.

Parents recognize the increased pressure on adolescents to engage in risky behaviors such as substance use (alcohol, cigarettes) and delinquent behaviors. Concern over substance use among adolescents in Thailand appears to be increasing as a result of recent indications of increases based upon reports from adolescents sampled from 8 secondary schools and 13 communities in Bangkok (2). The study estimated that 37% of Bangkok adolescents (mean age 15.5 years) overall have ever used alcohol, with 56.1% being classified as frequent drinkers. An increase in weekly and daily alcohol use among Thai secondary school students has also been reported (3). In addition, smoking rates have been reported as 11.6% in a school sample and 24.8% in a community sample of Bangkok adolescents (2). Researchers have called for effective programs for Thai adolescents out of concern about high usage rates (4).

Delinquency is also a concern. Among male and female youth surveyed in Bangkok, Thailand from 8 schools and 13 communities, 7.9% of the school sample and 11.2% of the community sample had carried a weapon and 29.8% of the school sample and 35.7% of the community sample had engaged in fighting (2). A total of 2.9% of the school sample and 2.6% of the community sample reported driving after drinking alcohol.

Prior research suggests that family processes and structures are particularly important for preventing the initiation of substance use (alcohol, cigarette, and other drugs) and for decreasing later misuse of substances in adolescents (5). Families play a key role in adolescent problem behavior through their role in the socialization of children to adapt to demands in their social context (5).

Parental monitoring includes tracking behaviors that regulate and provide awareness of a child's whereabouts, conduct, activities and companions (6). Low levels of parental monitoring are associated with high levels of youth problem behaviors in the U.S. e.g.,(7). There is reason to believe that monitoring may function in a similar fashion in Thailand. Relationships between monitoring and substance use and monitoring and delinquency have been broadly discussed across ethnicities and socioeconomic groups (7).

Parenting style can either contribute to, or prevent adolescent risks such as substance use and delinquency. Authoritative parenting style, characterized by high levels of parental control, as well as high levels of warmth/responsiveness, protects adolescents against both substance use and delinquency as compared to adolescents raised in non-authoritative families(8). In contrast, authoritarian parental practices, characterized by coercive and controlling parental behaviors, without warmth, increase risks of adolescent problem behaviors (9).

Adolescents who are close to parents are at increased likelihood to adopt conventional norms and values, thus protecting against deviant behavior (10).Conversely, weak family bonds are reported to be a risk factor for frequent and excessive drinking in adolescents (11). Parental closeness may be associated with better adolescent outcomes because the more time adolescents spend with parents and families who promote pro-social values, the fewer opportunities there are for adolescents to engage in substance use. In particular, maternal closeness is significantly related to protection against adolescent drinking (11). Although insufficient data exists on how closeness and bonding operates in Asian families, similar relationships may be expected to exist.

Parents can have a direct and important positive influence on their teen's alcohol and drug use through communication with their children (12). Parent communication was found to be a strong protective factor against drug use (12). For example, youth who perceived parental rules for alcohol and drug use had less drug use than those that didn't. Those rules were much clearer after a parent-child discussion although only 12-15% of the youth reported a discussion about drugs with their parent in the past year (12).

Family rituals (e.g., how families spend traditional holidays, special occasions) may also influence adolescent behaviors. (13). Rituals are a means by which the individual learns about culture and socialization. Rituals offer stability, structure and predictability and can act as a resource for coping with stress (14). Disruption of family rituals (e.g., parental fighting, drinking) may have a negative impact on adolescents and their behavior (13).

Experiences of parents in Thailand may be relevant for parents in other countries who are experiencing population migration from rural to urban communities. Further, these relationships between parenting practices and risky adolescent behaviors may be relevant to families immigrating to the western societies, thus demanding rapid adjustment of the family structure to the new environmental context.

Thai parents are faced with a rapidly changing cultural context in which they guide their adolescents. The Thai family culture contains much strength and these strengths may be retained even among families in urban Thai centers such as Bangkok. In contrast, the cultural changes and demands of a country rapidly moving towards an industrial and tourist based economy add pressures to the family systems. These pressures mirror other countries' experiences and experiences for immigrants in the United States.

In this paper, we address the prevalence of these risky behaviors for Thai adolescents. Our next focus is on the relationships between parenting and family practices and adolescent risky behaviors, specifically alcohol use, cigarette use and delinquency. Major parenting constructs are examined for their relative relationships to these risky adolescent behaviors, controlling for demographics: parental monitoring, parenting style, parental closeness, parental communication and family rituals. Understanding these relationships is relevant to understanding the best policies and practices for preventing adolescent risky behaviors for families with different cultural backgrounds and contexts.

Our study

Using a cross-sectional design, families (N=420) were randomly and proportionally selected from seven districts of Bangkok, Thailand. Sampling was based on the former Bangkok Metropolitan Administration (BMA) that divided districts into three zones (i.e., inner, middle, and outer zones). The population in each district (2006) was available from the Central Registration Bureau of the Department of Provincial Administration, Ministry of Interior. First, one district was selected from the inner zone, four from the middle zone, and two from the outer zone, using the probability proportional to size (PPS) sampling method (with case multiplication technique). Second, for each district, 35 blocks were selected using the PPS method of selection, resulting in a total of 245 blocks (35 blocks x 7 districts). This identified about 4,000 households in each district, or 30,471 households total across all seven districts, as target households. Third, block sampling selection was then carried out by the National Statistical Office (NSO) in collaboration with Mahidol University researchers. Using NSO maps of these 245 blocks, household census and enumerations in each block were conducted to identify eligible households (with adolescents aged 13-14 years old). Finally, based on this enumeration process, 60 households per district (420 households in total) were randomly selected to be interviewed. This represents 79.6% of eligible and contacted households participated.

Interview data were collected from one parent and one adolescent using an audio computer-assisted questionnaire (ACASI) on a laptop computer, for adolescent data collection. These findings are based only upon the adolescent reports for all measures except parenting style and household income which were attained from parent reports.

A total of 420 families with an adolescent between ages 13 and 14 were included in the sample. The adolescents were on average 13.45 years old (SD = .50) and 50.5% were female. The sample represented mostly ethnic Thais (91.2%), with 5.9% reporting their ethnicity as Thai-Chinese, and 2.9% reporting "Other ethnicity". The parents were predominantly female caregivers (83.1% female), and were aged 41.47 (SD = 6.35) on average. Most of the parents were married (82.6%). About three-fourths (74%) of the parents currently worked for pay, and 6% had graduated from college. Family monthly incomes ranged from less than 10,000 baht to more than 40,000 baht per month, currently equivalent to approximately $250 to $1,000 U.S. dollars. Specifically, 15.5% earned less than 10,000, 36.7% earned 10,001-20,000 baht, both categories which represent lower income for Bangkok families. Fifteen point two percent earned 20,001-30,000 baht, 6.5% earned 30,001-40,000, both of which represent middle income families. About a quarter (26.1%) of families earned over 40,000 baht, representing upper incomes.

Measures

To ensure measurement equivalency across the U.S. and Thai cultures, a number of steps were taken. First, interview instruments were developed through collaboration between Thai and U.S. research teams. Thai team members reviewed U.S. measures and provided interpretations of their cultural meaning, suggesting modifications, deletions, and additions. For example, questions regarding family rituals were modified so that answer choices reflected activities typical of Thai families during celebrations or special events. Second, the interview instruments were translated into Thai and then back-translated into English, using two different individuals to ensure an unbiased assessment. A careful review and comparison allowed further refinements where there was not an exact match. Finally, the instruments were reviewed by Thai parents/adolescents to provide feedback, prior to administering the final instrument to our sample. The following measures were included in the interview instruments.

Alcohol and cigarette use. Self-reports on adolescent alcohol use and cigarette use provided separate measures of lifetime use (1 = Yes, 0 = No).

Delinquency. Delinquency was measured using questions adapted from Elliott, et al. (15) and created into two indices: "serious" delinquency (6 items--participated in gang fights, gave drugs to friends, joined a gang at school, stopped by police and told to go home, taken to a police station and arrested, ran away from home) and "minor" delinquency (5 items--skipping school, shoplifting, joyriding, vandalized property, disorderly conduct).

Parental monitoring. Adapted from Patterson and Stouthamer-Loeber (16) and Capaldi and Patterson (17), two parental monitoring scales were created: parents' rule-setting behaviors and parents' knowledge of their adolescent's whereabouts. Youth responded to four items asking about their parents' rule and limit setting behaviors over the past 6 months, such as setting and enforcing curfews and restricting activities and companions (e.g., "how often did your parents really restrict where you were allowed to go?"). Youth answered three items regarding their parents' knowledge of their whereabouts and activities when away from home over the past 6 months (e.g., "how often did your parents really know what you were doing after school or when you were away from home?"). Responses for all items ranged from "none of the time" to "all/almost all of the time" on a 4-point scale. Items were averaged to create each scale (rules setting—Cronbach alpha =.68; knowledge of whereabouts—Cronbach alpha = .66).

Parental closeness. Parental closeness was measured using questions from the Add Health Study (18). Adolescents answered four questions, indicating both how close they felt to each parent, and how much they thought each parent cared for the adolescents themselves (e.g., "How close do you feel to your mother?". The youth answered on a 4-point scale ranging from "not at all" to "very much." The four items were averaged to create a closeness scale (Cronbach alpha = .66).

Parental communication. Parental communication was assessed through ten items adapted from Paschall, et al. (19). Youth reported on how often their parents talked to them about a variety of topics, such as plans for the day and drug use (e.g., How often has either of your parents talked with you about plans for the future?". Responses ranged from "never" to "most of the time/almost daily" on a 4-point scale. A scale was created by averaging the items (Cronbach alpha = .87).

Parenting style. Parenting style was based upon parent reports using questions adapted from the Parenting Styles and Dimensions Questionnaire (20). Parents answered how frequently they exhibited certain parenting behaviors, using a 4-point scale ranging from "never" to "most of the time/almost daily." The 36 items measured three types of parenting style: permissive, authoritarian, and authoritative parenting. Permissive parenting style was measured by five items (Cronbach's alpha = .64) asking the parent to rate the frequency with which they exhibited each of the five behaviors within an indulgent dimension, like "states punishments to the child and does not actually do them." Three dimensions, physical coercion, verbal hostility, and non-reasoning/punitive, measured authoritarian parenting style (Cronbach alpha = .82), with four questions each. For example, "Yells or shouts when child misbehaves" was included in the verbal hostility dimension. Authoritative parenting, which is theoretically the most positive parental approach, was measured by a scale (Cronbach alpha = .86) reflecting three dimensions, connection, regulation, and autonomy, each with four questions. For example, "explains consequences of behavior to child," was a behavior measured in the regulation dimension.

Family Rituals. Assessing Thai family culture and the important role that this cultural context may have on adolescent behaviors was addressed, in part by assessing family rituals. Questions were adapted from Fiese and Kline's Family Ritual Questionnaire (14), which are typical family behaviors or traditions at holidays, celebrations, or special events. Specifically, youth were asked to think of a special occasion typical of their family's celebrations that happened within the last 6 months. They chose from a list, answering yes (1) or no (0) to each, to denote activities their family engaged in for the event, such as "cook a special meal," and also reported whether they considered the event important. Two indices were created based on a priori conceptualizations of protection and risk according to item face value. The scale indicating protective rituals includes seven items: whether your family cooks a special meal, visits and pays respect to elders, "other" protective rituals the family participates in, whether the teen themselves participates in cooking the special meal with the family, the teen goes to the wat and/or makes merit, goes to a special ceremony, the teen visits and pays respect to elders, and whether the event was important to the teen. The scale indicating risks in rituals includes six items: participation by non-family members, family members drinking alcohol at these events, and four items regarding whether the teen is upset by some aspect of the family ritual (i.e., family fights, not being able to be with friends, is bored by the ceremony, or is upset by some other aspect).

Background variables. Teens reported their age and gender (1=male, 2=female), while parents reported their total household income (scale from 1=less than 10,000 baht to 5=over 40,000 baht).

Findings

Our three adolescent behaviors of interest, alcohol use, cigarette use, and delinquency, vary in the prevalence among Thai youth with alcohol use and delinquency the most common. About one-fifth (22.2%) of adolescents reported having ever used alcohol, with 9.3% having used in the past six months. Some form of delinquency was reported by 39% of the adolescents. Serious types of delinquency are reported by 28.2% of the adolescents. Specifically, 24.6% joined a gang at school, 14.7% participated in gang fights, 5.8% had been stopped by police and told to go home, 3.4% had run away from home (at least overnight), 2% gave drugs to friends, and 1.5% had been arrested. Minor types of delinquency are reported by 28.4% of the adolescents. Specifically, 17.1% had skipped school, 11.2% had engaged in disorderly conduct, 9.3% had shoplifted, 9.2% had engaged in vandalism, and 2.2% had gone joyriding. In contrast, cigarette use was less frequently reported by Bangkok youth. Lifetime use of cigarettes is reported by 9.6% and past six month use is reported by 5.5% of the sample.

Because gang involvement seemed high in prevalence, additional analyses were conducted on this construct. Specifically, we wanted to ensure that gang membership was interpreted by Thai youth to represent membership with youth who were involved in deviant activities. First, our international research team reviewed the translations and back translations of the instruments. We decided that the question was interpreted the same way in Thailand as in the U.S., the translation was accurate, and the meaning was the same (i.e., gang activity represent the same level of potentially serious behavior in Thailand as in the U.S.). We then subjected the construct to further analytic tests. Second, gang membership was positively and significantly correlated with all other types of serious and minor delinquency. Third, when we re-examined our delinquency data, excluding cases that had only gang membership and no other form of delinquency, similar results were found. Thus we concluded that we did have a meaningful construct that measured a delinquent behavior.

Parental practices and family rituals

Table 1. Mean scores on parental practices and family ritual measures

	Possible Range	*M* (*SD*) (N = 420)
Parental knowledge of whereabouts	0-4	3.05 (0.69)
Rules	0-4	2.82 (0.65)
Parental closeness	0-4	3.81 (0.39)
Parental communication	0-4	2.50 (0.76)
Permissive parenting style	0-4	2.06 (0.48)
Authoritarian parenting style	0-4	1.84 (0.42)
Authoritative parenting style	0-4	3.14 (0.46)
Protective family rituals	0-7	5.04 (1.76)
Ritual risk indicators	0-6	1.78 (1.28)

Table 1 presents the frequencies and distributions for parental practices and both protective family rituals and family ritual risk indicators. For questions pertaining to parental knowledge of their adolescent's whereabouts, family rules, parenting styles (authoritative, poor/inconsistent parenting), parental communications, and parental closeness, the response range is from a low of 1 to a high of 4. Adolescents report high levels of parental closeness (*M* = 3.81), authoritative parenting styles (*M* = 3.14) and knowledge of whereabouts (*M* = 3.05), with slightly less discussion of family rules (*M* = 2.82). In contrast to authoritative parenting style, parents are less likely to report themselves as using parenting styles characterized by permissive (*M* = 2.06) and authoritarian (*M* = 1.84). Adolescents report an average of 5 protective rituals from a possible 7 items. Adolescents report an average of 1.8 ritual risk indicators from a possible of 6 items.

Bi-variate correlations between adolescent and parenting behaviors

Correlations between parenting practices, family rituals and risky adolescent behaviors are presented in table 2. Higher levels of parental monitoring (i.e., knowledge of whereabouts) are associated with lower levels of adolescent cigarette use, minor and serious delinquency. Greater parental rule-setting is associated with lower rates of adolescent alcohol use and serious delinquency.

Parenting style is also related to adolescent behaviors. Higher levels of authoritative parenting style are associated with lower levels of adolescent cigarette use. Similarly, parental closeness is related to lower rates of cigarette use and both minor and serious delinquency. Higher levels of permissive parenting were associated with increased alcohol and cigarette use and both types of delinquency. Higher levels of authoritarian parenting style were associated with increased rates of minor and serious delinquency. Protective family rituals are associated with lower rates of adolescent cigarette use, minor and serious delinquency. Ritual risk indicators are related to higher rates of adolescent alcohol use, minor and serious delinquency. Our measures of parental communication show no relationship to adolescent risky behaviors.

Multivariate analyses

Multivariate analyses examine the relationship between parental monitoring (i.e., knowledge of whereabouts and rule-setting behaviors), parenting style, parental closeness, and family rituals with alcohol and cigarette use and delinquency, while controlling for individual demographic factors (e.g., gender and income). Parental communication is not included in multivariate analyses based on non-significant relationships in bi-variate correlations. We also examined possible interaction effects between gender and the risk and protective variables significantly associated with adolescent behaviors at the bi-variate level. Given that alcohol and cigarette use are categorical variables, the relative importance of the parental behaviors and adolescent demographics are examined using logistic regression analyses. Our analyses present main effects of these important variables.

Higher rates of ritual risk indicators (OR = 1.36; p < .01) are related to significantly greater prevalence of alcohol use, controlling for all other parental practices and adolescent demographics (see Table 3). No other variables are significantly related to adolescent alcohol use when entered simultaneously in this logistic regression.

Older adolescent age (OR = 2.42, p < .05), decreased knowledge of whereabouts (OR = .42; p < .01), and lower levels of authoritative parenting style (OR = .41, p < .05), are significantly related to higher prevalence of cigarette use, controlling for parental practices and other adolescent demographics (see Table 4).

Table 2. Bivariate correlations among key constructs

	1	2	3	4	5	6	7	8	9	10
1. Age										
2. Gender	.040									
3. Income	.005	.007								
4. Knowledge of whereabouts	-.037	.148†	-.026							
5. Rule-setting	-.005	.204†	.012	.275‡						
6. Authoritarian parenting style	-.068	-.034	.032	-.133†	.012					
7. Permissive parenting style	-.002	.049	.121*	-.191‡	-.055	.548‡				
8. Authoritative parenting style	-.062	-.025	.175‡	.072	.068	.191‡	.207‡			
9. Parental closeness	-.030	.007	-.006	.209‡	.126*	-.103*	-.104*	-.011		
10. Parental communication	-.029	.057	.080	.097*	.155†	-.057	-.017	.035	.114*	
11. Protective family rituals	-.120*	.097*	-.101*	.173‡	.033	-.049	-.037	.051	.249‡	.115*
12. Ritual risk indicators	.124*	.030	.041	-.040	.016	.045	.058	-.017	-.110*	.033
13. Alcohol use	.133†	-.022	.078	-.088	-.097*	.078	.119*	-.014	-.083	.021
14. Cigarette use	.120*	-.123*	-.047	-.226‡	-.035	.091	.111*	-.103*	-.131*	.030
15. Serious delinquency	.174‡	-.094	.038	-.256‡	-.126*	.121*	.124*	-.039	-.221‡	-.024
16. Minor delinquency	.123*	-.028	.054	-.182‡	-.048	.126*	.097*	.032	-.127*	-.067

$p \leq .05$, †$p < .01$, ‡$p < .001$.
Teen report for all variables except parenting style and household income.

Table 2. (Continued)

	11	12	13	14	15
12. Ritual risk indicators	.086				
13. Alcohol use	-.037	.200‡			
14. Cigarette use	-.141†	.081	.369‡		
15. Serious delinquency	-.235‡	.222‡	.333‡	.420‡	
16. Minor delinquency	-.163†	.191‡	.291‡	.385‡	.600‡

$^{*}p \leq .05$, $^{\dagger}p < .01$, $^{\ddagger}p < .001$.
\# Teen report for all variables except parenting style and household income

Table 3. Relationships for parental practices, family rituals with alcohol use #: Logistic regression

	Any Alcohol use vs. No Alcohol use (N = 378)		
	B	Exp(B)	95% CI
Teen age	0.48	1.62	0.97 - 2.71
Teen gender	0.04	1.04	0.62 - 1.75
Household income	0.12	1.12	0.94 - 1.34
Knowledge of whereabouts	-0.15	0.86	0.58 - 1.26
Rule-setting	-0.31	0.73	0.48 - 1.12
Authoritarian parenting style	0.12	1.13	0.55 - 2.30
Permissive parenting style	0.49	1.63	0.85 - 3.13
Authoritative parenting style	-0.15	0.86	0.48 - 1.53
Parental Closeness	0.01	1.01	0.52 - 1.97
Protective Family Rituals	-0.03	0.97	0.83 - 1.13
Ritual risk indicators	0.31	1.36**	1.12 1.65

\# Teen report for all variables except parenting style and household income.
** $p < .01$.

Using a Poisson regression, our model of risk and protective constructs is statistically significant for predicting serious delinquency (likelihood ratio chi-square = 124.37, df = 11, p < .001). We then conducted Poisson regression analyses including interaction terms for the interaction of gender with the other predictors. This model is also significant (likelihood ratio chi-square = 156.56, df = 19, p < .001). Comparing the change in chi-square and degrees of freedom between the two models, the model containing the interactions is a significantly better model than the model without interactions (chi-square = 32.20, df = 8, p < .001), so this model is presented in Table 5. There was a main effect for ritual risk indicators (B = -.34; p < .05) and this was in the opposite direction hypothesized with ritual risk indicators related to less delinquency but this relationship is made more understandable with the gender interactions. The main effects for adolescent gender and household income are not significantly related to serious delinquency. However, there is a significant interaction between adolescent gender and ritual risk indicators (B = .44, p < .001). Ritual risk indicators are related to increased serious delinquency for girls (B = .53, p < .001), but not boys (B = .11, p = .10).

Table 4. Relationships for parental practices, family rituals with cigarette use #: Logistic regression

	Any Cigarette use vs. No Cigarette use (N = 378)		
	B	*Exp(B)*	*95% CI*
Teen age	0.89	2.42*	1.10 - 5.36
Teen gender	-0.77	0.46	0.21 - 1.04
Household income	-0.13	0.88	0.67 - 1.16
Knowledge of whereabouts	-0.87	0.42**	0.24 - 0.73
Rule-setting	0.26	1.29	0.69 - 2.42
Authoritarian parenting style	-0.06	0.95	0.32 - 2.76
Permissive parenting style	0.93	2.53	0.94 6.79
Authoritative parenting style	-0.89	0.41*	0.17 - 1.00
Parental closeness	0.00	1.00	0.40 - 2.50
Protective Family Rituals	-0.08	0.92	0.74 - 1.15
Ritual Risk indicators	0.19	1.21	0.92 - 1.61

Teen report for all variables except parenting style and household income.
*p <.05, ** p < .01.

Our model of parenting practices and family rituals is also statistically significant for predicting minor delinquency (likelihood ratio chi-square = 65.26, df = 11, p < .001). A model containing interaction terms for the interaction of gender with the other predictors is also significant (likelihood ratio chi-square = 89.59, df = 19, p < .001). The difference in chi-square and degrees of freedom between the two models is significant (chi-square = 24.33, df = 8, p < .01), indicating that the model including the interactions was the better-fitting model, so this model is depicted in Table 6 and described below.

Table 5. Relationships for parental practices, family rituals with serious delinquency #: Poisson multiple regression

	B	SE	95% Wald Confidence Interval Lower	Upper	Wald χ^2
Teen age	0.57**	0.19	0.20	0.94	9.06
Teen gender	-3.11	1.67	-6.38	0.16	3.48
Household income	0.00	0.07	-0.14	0.14	0.00
Knowledge of whereabouts	-0.49	0.42	-1.31	0.33	1.35
Rule-setting	-0.54	0.50	-1.51	0.43	1.19
Authoritarian parenting style	0.41	0.79	-1.15	1.96	0.26
Permissive parenting style	-0.72	0.70	-2.09	0.66	1.04
Authoritative parenting style	-0.88	0.69	-2.23	0.47	1.64
Parental closeness	0.61	0.50	-0.37	1.59	1.48
Protective family rituals	-0.26	0.13	-0.53	0.00	3.82
Ritual risk indicators	-0.34*	0.16	-0.66	-0.03	4.56
Gender * Knowledge	0.07	0.27	-0.46	0.60	0.07
Gender * Rules	0.33	0.34	-0.33	0.99	0.94
Gender * Authoritarian parenting	-0.02	0.49	-0.98	0.93	0.00
Gender * Permissive parenting	0.57	0.46	-0.33	1.48	1.55
Gender * Authoritative	0.40	0.42	-0.42	1.22	0.92
Gender * Closeness	-0.63	0.33	-1.28	0.02	3.59
Gender * Protective rituals	0.12	0.09	-0.06	0.29	1.74
Gender * Risk indicators	0.44***	0.11	0.22	0.66	16.02
Omnibus likelihood ratio chi-square	**156.56*****				

\# Teen report for all variables except parenting style and household income.
* p < .05, *** p < .001.

The only significant main effect predictor of minor delinquency is rule-setting (B = -.94, p < .05). However, gender significantly interacts with two predictors, rule-setting and ritual risk indicators. Gender significantly interacts with rule setting (B = .65, p < .05) although it is only weakly and positively related to minor delinquency for girls at the trend level (B = .36, p = .08) and not significantly related to minor delinquency for boys (B = -.29, p = .14). In this case, the interaction indicates that the slopes for boys and girls are significantly different from one another, but neither significantly differs from zero. Interaction effects between gender and ritual risk indicators were significant (B = .41, p < .01) and positively correlated to minor delinquency for girls (B = .47, p < .001), but not for boys (B = .07, p = .37).

Table 6. Relationships for parental practices, family rituals with minor delinquency #: Poisson multiple regression

	B	SE	95% Wald Confidence Interval Lower	Upper	Wald χ^2
Teen age	0.33	0.19	-0.03	0.70	3.17
Teen gender	-1.42	2.07	-5.48	2.65	0.47
Household income	0.06	0.07	-0.07	0.19	0.74
Knowledge of whereabouts	-0.34	0.48	-1.28	0.60	0.50
Rule-setting	-0.94*	0.44	-1.80	-0.08	4.61
Authoritarian parenting style	1.14	0.68	-0.18	2.47	2.85
Permissive parenting style	0.00	0.70	-1.38	1.37	0.00
Authoritative parenting style	-0.39	0.75	-1.86	1.09	0.26
Parental closeness	0.61	0.64	-0.65	1.86	0.90
Protective family rituals	-0.20	0.16	-0.52	0.11	1.67
Risk indicators	-0.34	0.19	-0.71	0.02	3.39
Gender * Knowledge	0.01	0.31	-0.60	0.62	0.00
Gender * Rules	0.65*	0.28	0.11	1.20	5.49
Gender * Authoritarian parenting	-0.40	0.46	-1.31	0.50	0.76
Gender * Permissive	-0.08	0.47	-1.00	0.85	0.03
Gender * Authoritative	0.33	0.46	-0.57	1.22	0.52
Gender * Closeness	-0.53	0.38	-1.29	0.22	1.93
Gender * Protective rituals	0.09	0.10	-0.10	0.29	0.92
Gender * Risk indicators	0.41**	0.12	0.17	0.65	11.28
Omnibus likelihood ratio chi-square	**89.59***				

\# Teen report for all variables except parenting style and household income.
*p < .05, ** p < .01, *** p < .001.

Discussion

Rates of adolescent risky behaviors in Bangkok, Thailand were similar to rates for adolescents in the United States. (21). Multi-variate analyses show that one of the more consistent constructs related to three of the four adolescent risk behaviors under investigation was family ritual risk indicators. Particularly for girls, higher levels of ritual risk indicators are related to more adolescent risky behaviors. Knowledge of whereabouts was related to less adolescent cigarette use and rule setting was related to less minor delinquency for girls but not for boys. These analyses suggest that prevention of alcohol use, cigarette use and delinquency in Thailand, will follow similar models to those found in the US. For example, parental monitoring (knowing the whereabouts of your child) and parenting style are relevant to Thai adolescent behaviors. Although not as well studied in the US, protective rituals and ritual risk indicators are important to adolescent risky behaviors in Thailand, particularly for females and may assist us in understanding the interaction of parenting and cultural influences in Thailand.

These data, due to their cross-sectional nature, do not allow us to determine causal pathways between these parenting behaviors and adolescent risky behaviors. Our study was all within Bangkok and different relationships between parenting practices and adolescent risk behaviors may exist in rural areas outside of Bangkok. Our measures of family culture are limited, albeit important indicators. The self-reported levels of current (last six month) alcohol and cigarette use are low, and parental influences may be more evident as greater variation in ATOD use emerges as youth become older.

Measures of Thai family culture in the form of family rituals suggests that future studies should seek to explore more carefully, how family values, traditions, and expectations are transmitted across the generations. Bangkok families are undergoing stresses of urban lifestyle and influences of Western cultures. Despite the more homogeneous cultural backgrounds Thais share as compared to many other nations, differences in parenting practices and family rituals are noted. These differences are related to increased or decreased levels of adolescent risky behaviors. This suggests that cultural contexts are not totally influencing adolescent behaviors and differences in parenting styles and family rituals are still important.

Thai family influences on young males and females need further exploration. The stronger influence of family rituals for girls may be related to girls being more influenced by the family milieu. Recent findings from Australian middle school students suggest that the parental influences on girls may differ from boys with regards to drinking behavior (22). One measure of the family functioning and differences for girls and boys may be in family rituals. Family rituals serve to reflect the overall organization of the family as a unit. Families that are less organized and cohesive have more disrupted family rituals, and this may impact girls more than boys.

The influence of same sex parenting vs. opposite sex parenting needs further exploration for the Thai culture. Although women's roles in Thailand are reportedly changing (1), the parenting practices of mothers may be most important to their daughters and mothers may still not have an equal influence on their sons behaviors. If mothers are the primary parent for controlling adolescent behaviors (e.g, monitoring, connecting, setting rules), they may have less authority over their male, as compared to female, adolescents. These questions deserve further exploration.

These findings suggest that prevention programs and strategies using parenting practices to prevent adolescent alcohol use, cigarette use and other delinquent behaviors are appropriate for the Thai culture. Our findings indicate that the majority of youth have not initiated alcohol use at ages 13-14, and thus, this age range would be appropriate for the delivery of a prevention programs or strategies. Such programs need to be developed or adapted for the Thai family culture. In particular, the findings on family rituals, both as a positive and negative influence on youth, suggest one way in which adaptation may be guided. Further research is needed to determine how the Thai families may implement and practice these constructs and additional strengths in the Thai culture that may also serve to protect Thai youth from problem behaviors.

Acknowledgment

Research for and preparation of this manuscript was supported by NIAAA 1R01AA015672-01A1 "Youth Alcohol Use and Risky Sexual Behavior in Bangkok". Content is the sole responsibility of the authors and does not necessarily represent the official views of NIAAA.

References

[1]   Pinyuchon M, Gray LA, House RM. The Pa Sook Model of counseling Thai families: A culturally mindful approach. J Fam Psychother 2003;14(3):67-93.

[2]   Ruangkanchanasetr S, Plitponkarnpim A, Hetrakul P, Kongsakon R. Youth risk behavior survey: Bangkok, Thailand. J Adolesc Health 2005;36(3):227.

[3]   Daosodsai P, Bellis MA, Hughes K, Hughes S, Daosodsai S, Syed Q. Thai War on Drugs: Measuring changes in methamphetamine and other substance use by school students through matched cross sectional surveys. Addict Behav 2007;32(8):1733.

[4]   Assanangkornchai S, Pattanasattayawong U, Samangsri N, Mukthong A. Substance use among high-school students in southern Thailand: Trends over 3 years (2002-2004). Drug Alcohol Depend 2007;86(2-3):167.

[5]   Velleman RD, Templeton LJ, Copello AG. The role of the family in preventing and intervening with substance use and misuse: a comprehensive review of family interventions, with a focus on young people. Drug Alcohol Rev 2005;24(2):93-109.

[6]   Dishion TJ, McMahon RJ. Parental Monitoring and the Prevention of Child and Adolescent Problem Behavior: A Conceptual and Empirical Formulation. Clin Child Fam Psychol Rev 1998;1(1):61.

[7]   Tragesser SL, Beauvais F, Swaim RC, Edwards RW, Oetting ER. Parental monitoring, peer drug involvement, and marijuana use across three ethnicities. J Cross-Cultural Psychol 2007;38(6):670.

[8]   Newman K, Harrison L, Dashiff C, Davies S. Relationships between parenting styles and risk behaviors in adolescent health: an integrative literature review. Revista Latino-Americana De Enfermagem 2008;16(1):142.

[9]   Adalbjarnardottir S, Hafsteinsson LG. Adolescents' perceived parenting styles and their substance use: Concurrent and longitudinal analyses. J Res Adolesc 2001;11(4):401.

[10] Bell NJ, Forthun LF, Sun SW. Attachment, adolescent competencies, and substance use: Developmental considerations in the study of risk behaviors. Subst Use Misuse 2000;35(9):1177-206.

[11] Zhang L, Welte JW, Wieczorek WF. The influence of parental drinking and closeness on adolescent drinking. J Stud Alcohol 1999;60(2):245.

[12] Kelly KJ, Comello MLG, Hunn LCP. Parent-child communication, perceived sanctions against drug use, and youth drug involvement. Adolescence 2002;37:775-87.

[13] Fiese BH, Tomcho TJ, Douglas M, Josephs K, Poltrock S, Baker T. A review of 50 years of research on naturally occurring family routines and rituals: Cause for celebration? J Fam Psychol 2002;16(4):381.

[14] Fiese B, Kline CA. Development of the family ritual questionnaire: Initial reliability and validation studies. J Fam Psychol 1993;6(3):290-9.

[15] Elliott D, Ageton S, Huizinga D, Knowles B, Canter R. The prevalence and incidence of delinquent behavior: 1976-1980. Boulder, CO: Behavioral Research Institute, 1983.

[16] Patterson GR, Stouthamer-Loeber M. The correlation of family management practices and delinquency. Child Dev 1984;55:1299-307.

[17] Capaldi D, Patterson GR. Psychometric properties of fourteen latent constructs from the Oregon Youth Study. New York: Springer, 1989.

[18] Harris K, Mullan F, Florey J, Tabor PS, Bearman J, Udry JR, et al. The National Longitudinal Study of Adolescent Health: Research design. 2003 [updated 2003; cited]; Available from: . http://www.cpc.unc.edu/projects/addhealth/design.

[19] Paschall MJ, Ringwalt CL, Flewelling RL. Effects of parenting, father absence, and affiliation with delinquent peers on delinquent behavior among African-American male adolescents. Adolescence 2003;38(149):15-34.

[20] Robinson CC, Mandleco B, Olsen SF, Hart CH. The parenting styles and dimensions questionnaire (PSDQ). In: Perlmutter BF, Touliatos J, Holden GW, eds. Handbook of family measurement techniques: Vol 3 Instruments and index. Thousand Oaks, CA: Sage, 2001:319-21.

[21]   2007 National Survey on Drug Use and Health: Detailed Tables: Substance Abuse and Mental Health Services Administration-Office of Applied Studies, 2007|.

[22]   Kelly AB, Toumbourou JW, O'Flaherty M, Patton GC, Homel R, Connor JP, et al. Family relationship quality and early alcohol use: Evidence for gender-specific risk processes. J Stud Alcohol Drugs 2011;72(3):399-407.

In: Public Health Concern   ISBN: 978-1-62948-424-2
Editors: J Merrick and A Tenenbaum   © 2014 Nova Science Publishers, Inc.

Chapter V

Smoking from early to late adolescence in Greece

Evdoxia Kosmidou, PhD, Mary Hassandra, PhD[*],
Antonis Hatzigeorgiadis, PhD, Marios Goudas, PhD
and Yiannis Theodorakis, PhD
Department of Physical Education and Sport Science,
University of Thessaly, Karies, Trikala, Greece

Outcome expectancies are an individual's anticipation of a systematic relationship between events in some future situation and can either promote or inhibit smoking behavior. This chapter investigated differences between Greek adolescents' age levels in smoking outcome expectancies and assessed the impact of outcome expectancies on the likelihood that adolescents would be smokers or not. Study group: Participants were 655 Greek adolescents (311 males – 344 females) with an average age of 17.28 and they were assigned in 3 groups: the early adolescent group (11-14), the mid adolescent group (15-18) and the late adolescent group (19-22). Methods: Adolescents completed self reported questionnaires assessing their smoking behavior and their smoking outcome expectancies. Results: The chi-square test showed that there was a medium association between age level and smoking status. There were significant differences between the three age groups on general costs,

[*] Correspondence: Hassandra Mary, Department of Physical Education and Sport Science, University of Thessaly, Karies, 42100, Trikala, Greece. E-mail: mxasad@pe.uth.gr.

social costs, health costs and affect control. The investigation of the differences on each variable revealed that only the early adolescent group was significantly different from middle and late adolescent groups. Three logistic regressions assessed the impact of outcome expectancies on the likelihood that adolescents would be smokers or not. The only common predictor of being a smoker in all age levels, controlling for all other factors in the model, was the affect control variable. Conclusions: Smoking outcome expectancies differs across adolescents' age levels and contributes differentially to the classification of the participants as smokers and non-smokers. These differences must be taken into account when intervention programs are designed.

Introduction

Health risks from early smoking initiation are severe and unhealthy adolescents' behavioral patterns often extend into adulthood (1). According to a recent study in Europeans (2), while smoking among adults has declined, adolescent's smoking remains unchanged. Smoking is a serious problem among Greek youth (3), as the percentage of non-smoker students (13-15 years old) is steadily declining (4).

Until recently, the insufficient antismoking policy and the lack of educational antismoking programs, particularly in schools, were some of the possible causes of these disappointing statistics. The distribution of smoking status by adolescents' age is increasing from early to late adolescence. As Giannakopoulos et al (5) stated, smoking prevalence shows an increasing trend by age, beginning from 0.6% at the age of 12 and reaching 25.2% at the age of 17 years. The same increase was detected in an earlier study of Francis et al (3) who reported a sharp escalation of smoking from junior to senior high school.

Adolescence is characterized as a time of experimentation for many health-related and risk-taking behaviors. Adolescents' outcome expectancies, typically defined as an individual's anticipation of a systematic relationship between events in some future situation (6), can either promote or inhibit health related behavior. There is a considerable body of research examining outcome expectancies of health risk behaviors, like alcohol (7), drug use and risky sexual behavior (8), and smoking (9). Expectancies about smoking include perceived benefits or negative consequences one might anticipate from smoking. It appears that smoking expectancies are formed in childhood prior to personal experience with smoking and can predict the initiation of smoking

(10). Generally, research has shown that individuals who expect positive social and physiological consequences from smoking (e.g., social facilitation, relaxation, mood enhancement) are more likely to initiate and continue smoking than individuals who expect negative (e.g., health consequences) or less positive consequences (5). There is also evidence suggesting that heavier and more addicted smokers tend to hold more positive expectancies about the consequences of smoking than nonsmokers or lighter smokers (11). In particular for adolescents, it has been reported that expectancies predict the level of intention to smoke (12), which in turn predicts whether never-smokers would move to experimentation, as well as whether experimenters move to regular smoking (13). Among the expectancies factors that have been examined as predictors of tobacco use during adolescence, control of negative affect and weight management have been reported to significantly mediate the influence of current smoking on future smoking (14). In addition, general costs, social costs, and affect control were significant predictors of current smoking in a study by Hine et al (15).

Individual outcome expectancies are not stable over time. Among the factors that have been considered in relation to the shaping of smoking outcome expectancies in adolescents is age and experimentation or smoking status, with the latter appearing as the most powerful factor. A recent study of Urban and Demetrovics (16) found that age did not predict directly the expectancies variables of negative consequences, positive reinforcement, negative reinforcement and appetite–weight control for 953 high-school students of 14–20 years old. Another recent study by Chung et al (17) examined the development of positive smoking expectancies and smoking behavior in an urban cohort of girls followed annually over ages 11–14. It was reported that positive smoking expectancies were relatively stable over ages 11–14, despite an overall increase in the prevalence of past year smoking.

Wahl et al (18) detected no significant age differences on expectancy scores for students ranged in age from 14 to 19 years, but they found that changes in smoking expectancies were consistent with changes in smoking behavior over time. According to Anderson, Pollak and Wetter (19), expected outcomes of smoking were related to current smoking, experimentation, and susceptibility among never-smokers, even after controlling for key correlates of smoking behavior, including gender, grade, ethnicity, and peer smoking. Although more negative than positive smoking outcomes were accessible from memory, more positive than negative expected outcomes were correlated with smoking behavior.

In conclusion, the variation of outcome expectancies according to adolescents' age and the interpretation of adolescents' smoking status through their smoking outcome expectancies need further investigation. It is therefore important to examine adolescents' expectancies as they can either promote or inhibit behavioral responses to certain events. Considering the prevalence of youth smoking in the Greek population, the purpose of the present study was a) to examine how smoking outcome expectancies vary as a function of age and smoking status, and b) to assess the relation between smoking outcome expectancies and the likelihood of being smoker. It was hypothesized that:

- Participants in the late adolescence would smoke more cigarettes than participants in the early adolescence.
- There will be significant differences on smoking outcome expectancies by age, with an increase on positive outcome expectancies and decrease in negative outcome expectancies, from early to late adolescence.
- Smoking outcome expectancies would be significantly classify adolescents as smokers or non-smokers in all age levels, with the dimensions of affect control, general and social costs, and weight management, having the stronger relationships.

Our study

Participants were 655 Greek adolescents (311 males – 344 females) from 6 different cities across Greece, with an average age of 17.28 years (SD: 3.58, range: 11–22). Participants were divided into 3 groups according to their age. The early adolescence group (11-14 years old) comprising 30.7% of participants (106 boys and 95 girls); the mid-adolescence group (15-18 years old) comprising 36.3% of participants (114 boys, 124 girls); and the late adolescence group (19-22 years old) comprising 33.0% of participants (91 boys, 125 girls).

Smoking behavior was assessed with the quantity measure of smoking by Theodorakis & Hassandra (20). In particular, participants were asked to indicate how many cigarettes they smoked the day before on a 7-point scale: (a) 0, (b) 1–5, (c) 6–10, (d) 11–15, (e) 16–20, (f) 21–25, and (g) 26 + cigarettes per day.

Smoking outcome expectancies were assessed using the Adolescent Smoking Expectancies Questionnaire (ASEQ)(15). The original questionnaire comprises 6 scales (32 items) labeled: "General Costs" (10 items), "Social Benefits" (6 items), "Social Costs" (4 items), "Health Costs" (8 items), "Weight Control" (4 items), and "Affect Control" (4 items). Participants are asked to indicate their beliefs about the likelihood that the listed outcome will happen to them personally if they smoked cigarettes (1 _ very unlikely, 2 _ moderately unlikely, 3 _moderately likely and 4 _ very likely).

The ASEQ was adapted in the Greek adolescent population and the factor structure and psychometric properties of the smoking expectancy measure were reexamined. The items were translated into the Greek language by two bilingual health researchers. Then the reverse procedure was followed and inconsistencies were spotted and resolved by mutual agreement. A factor principle axis analysis was conducted with all 32 items. Oblimin rotation was used. Purpose was to identify and compute composite scores for the factors underlying the Greek version of the ASEQ. The initial Eigenvalues showed that the six factors explained 65.9% of the variance. During several steps, two items from the scale general costs were eliminated ("damage health of others" and "less spending money") because they did not contribute to a simple factor structure and failed to meet minimum criteria of having a primary factor loading of .4 or above. Additionally the items of the general costs scale "become dependent on nicotine" and "get hooked" had a factor loading above .40 on the health costs scale and no cross loadings with other scales, whereas, the item "hurt lungs" had a primary factor loading of .55 on the health costs scale (factor) and a cross-loading of .41 on general costs. These three items were relocated to the scale of health costs because they were very close in meaning to this scale (factor) also. A principle-axis factor analysis of the remaining 30 items, using oblimin rotation was conducted, with the six factors explaining 67.8% of the variance. The final version of the Greek version of the ASEQ consisted of 30 items, representing the scales: "General Costs" (5 items), "Social Benefits" (6 items), "Social Costs" (4 items), "Health Costs" (7 items), "Weight Control" (4 items) and "Affect Control" (4 items). Overall, the analyses indicated six distinct factors underlying adolescent responses to the Greek version of the ASEQ.

The study was approved by the ethics committee of the university. Permission to participate in the study was granted by adolescents' guardians, except for the late adolescent group who provided consent themselves. Participants were informed that their responses would be treated confidentially, that participation was voluntary, and that they could withdraw at any time without incurring any penalty. Questionnaires were completed in school settings, during class time, in the presence of at least one researcher and the classroom teacher. The completion lasted between 10 and 15 minutes.

Relation between adolescent's age level and current smoking status was examined using Chi-square test for independence. Group differences were examined using a multivariate analysis of variance. Finally, the degree to which smoking outcome expectancies would correctly classify adolescents as smokers or non-smokers was examined by conducting a logistic regression for each age level.

Findings

A Chi-square test for independence (with Yates Continuity Correction) indicated significant relation between adolescents' age level and current smoking status ?2 (1, n = 654) = 80.53, p < .001, phi = .35, with a marked increase in smokers as age increases (see table 1).

Table 1. Percentages of current smoking status according to adolescents' age level

	Current smoking status	
	Non smokers	Smokers
Early adolescents (11-14 years old)	92.5%	7.5%
Mid adolescents (15-18 years old)	69.7%	30.3%
Late adolescents (19-22 years old)	52.8%	47.2%

Table 2. Correlations, Means and Cronbach's alpha for the outcome expectancies variables

	Correlations					Mean (S.D.)	Cronbach's alpha
1. General costs	$-.154^{**}$	$.257^{**}$	$.700^{**}$	$.127^{**}$	$-.053$	3.05 (.77)	.90
2. Social benefits		$.218^{**}$	$-.184^{**}$	$.244^{**}$	$.272^{**}$	1.86 (.73)	.86
3. Social costs			$.259^{**}$	$.223^{**}$	$-.065$	1.97 (.83)	.81
4. Health costs				$.168^{**}$	$-.003$	3.31 (.74)	.91
5. Weight control					$.387^{**}$	2.39 (.84)	.84
6. Affect control						2.38 (.88)	.87

**. Correlation is significant at the 0.01 level (2-tailed).

Descriptive statistics, correlations, and Cronbach's alpha for the Outcome Expectancies variables are presented in table 2. The highest outcome expectancies scores were detected on general costs, weight control and affect control.

Group differences

A one-way MANOVA was calculated to test for differences in smoking outcome expectancies as a function of age group. Tests of normality, linearity, univariate and multivariate outliers, homogeneity of variance-covariance matrices, and multicollinearity, showed no violations of the assumptions. The analysis revealed a statistically significant multivariate effect, F (12, 1280) = 18.47, p = .000; Wilks' Lambda = .73; η^2 = .14. Examination of the univariate effects revealed significant differences for general costs, F (2, 645) = 20.91, p = .000, η^2= .06; social costs, F (2, 645) = 69.28, p = .000, η^2 =.17; health costs, F (2, 645) = 3.65, p = .026, ?2 =.01; and affect control, F (2, 645) = 23.65, p = .000, ?2 =.06, whereas no significant differences emerged for social benefits, F (2, 645) = 2.80, p = .061, η^2 =.009; and weight control F (2, 645) = .56, p = .570, η^2 =.002. Scheffe's post-hoc tests showed that the early adolescence group scored higher on general costs, social costs and health costs, and lower on affect control than the mid- and late adolescence groups. Mean scores for the three age groups are presented in table 3.

Table 3. Outcome expectancies scores according to adolescents' age level

Outcome expectancies	Adolescent level	N	Mean	SD
General costs	Early adolescents	197	3.37	.82
	Mid adolescents	236	2.87	.91
	Late adolescents	215	2.96	.79
Social costs	Early adolescents	197	2.50	.87
	Mid adolescents	236	1.76	.72
	Late adolescents	215	1.71	.68
Health costs	Early adolescents	197	3.43	.77
	Mid adolescents	236	3.25	.77
	Late adolescents	215	3.27	.65
Affect control	Early adolescents	197	2.05	.91
	Mid adolescents	236	2.45	.86
	Late adolescents	215	2.62	.79

Regressions

Three logistic regressions (one for each age group) were performed to assess the degree to which smoking outcome expectancies would correctly classify adolescents as smokers or non-smokers (table 4). Each model contained six independent variables (general costs, social benefits, social costs, health costs, weight control and affect control), whereas the dependent variable was adolescents' smoking status. Smoking status variable was obtained from current smoking status where 0 was maintained as non smokers and all the other values were considered as 1 (smokers).

Table 4. Logistic regression predicting likelihood of smoking status during adolescents' age level

	B	S.E.	Wald	df	p	Odds Ratio	95% C.I. for Odds Ratio Lower	Upper
Early adolescents								
General costs	-.22	.55	.16	1	.685	.79	.27	2.36
Social benefits	.57	.41	1.92	1	.166	1.77	.78	4.01
Social costs	-.93	.46	4.06	1	.044	.39	.16	.97
Health costs	-.19	.58	.11	1	.740	.82	.26	2.59
Weight control	.17	.44	.15	1	.692	1.19	.50	2.83
Affect control	1.25	.42	8.63	1	.003	3.51	1.52	8.13
Mid adolescents								
General costs	-1.25	.26	22.83	1	.000	.28	.17	.47
Social benefits	.53	.28	3.50	1	.414	.81	.50	1.32
Social costs	-.20	.24	.66	1	.000	.36	.20	.64
Health costs	-1.00	.28	12.30	1	.061	1.71	.97	3.01
Weight control	-.01	.20	.00	1	.931	.98	.65	1.47
Affect control	.97	.22	18.58	1	.000	2.65	1.69	4.11
Late adolescents								
General costs	-.42	.24	2.99	1	.083	.65	.40	1.05
Social benefits	.19	.28	.46	1	.497	1.21	.69	2.09
Social costs	-.35	.25	1.87	1	.171	.70	.42	1.16
Health costs	-.52	.29	3.13	1	.077	.59	.33	1.05
Weight control	-.19	.23	.68	1	.408	.82	.52	1.29
Affect control	.98	.23	17.13	1	.000	2.67	1.68	4.26

For the early adolescence group, the full model containing all predictors was statistically significant, χ^2 (6, N = 197) = 30.55, p = .000, indicating that the model could distinguish between smoking and non-smoking adolescents. The model explained between 14.4% (Cox and Snell R square) and 35.8% (Nagelkerke R squared) of the variance in smoking status, and correctly classified 92.9% of cases. Two of the six independent variables made a unique statistically significant contribution to the model (social costs and affect control). The strongest contributor of smoking was affect control, recording an odds ratio of 3.51. This indicated that, controlling for all other factors in the model, those expecting that smoking could have affect control as an outcome were over 3.5 times more likely to be smokers than those who did not expect this. The respective odds ratio for social costs was .39.

For the mid adolescence group, the full model containing all predictors was statistically significant, χ^2 (6, N = 236) = 63.06, p = .000, indicating that the model could distinguish between smoking and non-smoking adolescents. The model as a whole explained between 23.4% (Cox and Snell R square) and 33.1% (Nagelkerke R squared) of the variance in smoking status, and correctly classified 75% of cases. Three of the six independent variables made a unique statistically significant contribution to the model (general costs, social costs and affect control). The strongest contributor of being a smoker was affect control, recording an odds ratio of 2.64. The odds ratio for social costs and general costs were .36 and .29 respectively.

For the late adolescence age level, the full model containing all predictors was statistically significant, χ^2 (6, N = 119) = 37.25, p = .000, indicating that the model could distinguish between smoking and non-smoking adolescents. The model as a whole explained between 15.9% (Cox and Snell R square) and 21.2% (Nagelkerke R squared) of the variance in smoking status, and correctly classified 64.2% of cases. Affect control was the only independent variable making a unique statistically significant contribution to the model, recording an odds ratio of 2.67.

Discussion

The purpose of the present chapter was to examine smoking behavior and smoking outcome expectancies as a function of age, and also to assess the degree to which outcome expectancies could predict smoking status at the different age groups. Overall, the results revealed a significant association between age group and smoking status; also that early adolescents have different general costs, social costs, health costs, and affect control outcome expectancies; finally that expectancies could predict smoking status across all age groups.

The increase of Greek smokers from early to late adolescence is in line with the existing findings (3,4) and was described by a moderate association between age group and smoking status. The percentage of smokers of the late adolescence group attains the percentages of the Greek adults' smoking percentage (21).

The scores on the outcome expectancies dimensions of general, social and health costs were lower for the early compared to the mid- and late adolescence groups. This implies that older children expect fewer negative consequences (general, social and health) and this may be attributed to several factors like, the socializing with other peer smokers (e.g.: social factors), first hand experiences (e.g.: general factors), or the expected time of occurrence of these negative outcomes (health factors). In contrast, scores on affect control were lower in the early adolescence group, suggesting that older adolescents expected that smoking had more control power on their affective conditions than younger adolescence. A possible explanation could be that at later ages smoking experiences have initiated which may influence affective responses and increase affective dependence. The hypotheses regarding weight control and social benefits were not verified on this sample. Regarding weight management it could be argued that it may not be a prevailing issue at these age levels and thus did not differentiate between the adolescence groups of this study.

In the logistic regression analyses the percentages of the variance in smoking status explained by the model and the correctly classified cases was generally high, yet it should be noticed that it decreased from early to late adolescence, suggesting that for older adolescents the present measured outcome expectancies had less power to classifying them as smokers or nonsmokers. Probably, a broader range of expected outcomes but also other variables interfere in differentiating smokers from nonsmokers. In a future study more outcome expectancies, like addiction, financial costs, physical appearance, and boredom reduction, could be included to test if the classification power of the outcome expectancies models could be improved.

The only outcome expectancy factor that predicted smoking behavior in all age groups was affect control. In addition, high odds ratios were recorded for affect control, suggesting that this is the most influential factor distinguishing smokers from nonsmokers regardless of age. Similar results have been reported for adolescents with a mean age of 14 years (22). In that study adolescents believed that smoking reduced anxiety or negative affect regardless of smoking status. The results of the present study seem to support and extend this finding for all adolescence stages. Several studies have highlighted the impact of affect regulation on smoking behavior. Management of negative affective states, such as depression, anxiety, and anger is a strong motive and predictor of smoking behavior. Cohen et al (23) examined whether the tendency to experience negative affective states and smoking outcome expectancies predict smoking behavior over time in young adults. Their results indicated that dispositional negative affect and positive smoking expectancies were significantly correlated with smoking behavior both within and across time. The social costs variable was the second strongest outcome expectancy factor.

The lack of contribution of the weight control and health costs factors is also an interesting finding. Weight control has given contradictory results in adolescents' population as a contributor to smoking behavior (14, 15). Nevertheless, in the present study weight control had no power to classify participants as smokers or non-smokers.

Health costs also had no power to classify adolescents' smoking condition. Both smokers and non-smokers adolescents are aware of the future possible negative health consequences of smoking, but this awareness does not seem to play important role in their smoking behavior. Knowledge about tobacco among adolescents was essentially not related with future tobacco use according to a study with 2,581 adolescent students (24). This is in line with recent directions on health education, which suggest that knowledge does not necessary affect beliefs, attitudes, intent and behavior (25). The above findings suggest that the affect control and the social costs factors should be addressed in smoking prevention programs and that these two smoking related issues should be the focus in Greek adolescent smoking cessation programs.

In general the present results are in accordance with the prior work of Hine et al (12), who found that three of the six subscales examined in their study, general costs, social costs, and affect control, were significant predictors of current smoking. In addition, the present study showed that these subscales contribute differentially to the classification of adolescents as smokers – non-smokers across age level. This information can be valuable when smoking cessation programs are designed.

A few limitations are important to consider when interpreting these findings. First, the sampling procedure did not assure representativeness of adolescents in Greece, although participants were from different cities, and therefore, no attempt at generalizing the findings should be made. Also, the study relied on self-report measures of adolescents' smoking behavior which was not biochemically verified. Self reported measures of behavior have been criticized as liable to social desirability. However, self-report is a source of information that is inaccessible to others and is a method that is difficult to replace when assessing large samples. Additionally, for measuring subjective beliefs such as expectancies, surveys are the most feasible option. Among the weaknesses of the present work is the cross-sectional nature of the study, which does not allow for any causal inferences to be made. Nevertheless, cross-sectional studies reveal associations that may guide further research and provide ground for generating hypotheses.

Finally, as many researchers postulate there is a reciprocal causation in the relationship between expectancies and behavior (26). According to the present results older adolescents show different smoking outcome expectancies compared to younger, but we cannot attribute this only to age related factors, such as differences in the cognitive decision, because smoking behavior also increased at the same time. Thus, it is also possible that the flow of causality runs from smoking to expectancies rather than from expectancies to smoking. Therefore, results of the present study must be cautiously encountered, with regard the causality between outcome expectancies and smoking status.

Future research

Future studies are needed to explore the relationship between changes in smoking expectancies and smoking behavior over time. Longitudinal designs, experimental studies and larger samples of adolescents are needed to assess more directly the meanings of the present findings. Also, a broader range of smoking related outcome expectancies dimensions may be useful in order to better understand the relationships between expectancies and smoking behavior. Additional factors, both interpersonal and intrapersonal, and several unaccounted factors (such as culture, ethnicity, and the access to and use of other tobacco products and illicit substances) may affect adolescents' smoking behaviors and could also be taken into account in future endeavors. Finally, qualitative investigation of how negative expectancies turn to less negative or positive expectancies should be a future direction, for the better understanding of the factors and the mechanisms that regulate this complex and dynamic procedure.

In conclusion, the present findings suggest that smoking expectancies differ across adolescence stages and contribute differentially to the classification of the participants as smokers or non-smokers.

References

[1]   Paavola M, Vartianinen E, Puska P. Predicting adults smoking: the influence if smoking during adolescence and smoking among friends and family. Health Edus Res 1996;11(3):309-15.

[2]   De Vries H, Mudde A, Leijs I, Charlton A, Vartiainen E, Buijs G, et al. The European Smoking prevention Framework Approach (EFSA): an example of integral prevention. Health Educ Res 2003; 18(5):611-26.

[3]   Francis K, Katsani G, Sotiropoulou X, Roussos A, Roussos C. Cigarette smoking among Greek adolescents: Behavior, attitudes, risk, and preventive factors. Subst Use Misuse 2007;42:1323-36.

[4]   Kyrlesi A, Soteriades ES, Warren CW, Kremastinou J, Papastergiou P, Jones NR, Hadjichristodoulou C. Tobacco use among students aged 13-15 years in Greece: the GYTS project. BMC Public Health 2007;7:3.

[5]   Giannakopoulos G, Panagiotakos D, Mihas C,Tountas Y. Adolescent smoking and health-related behaviours: interrelations in a Greek school-based sample. Child Care Health Dev 2009;35:164-70.

[6]   Goldman MS, Brown SA, Christiansen BA. Expectancy theory: Thinking about drinking. In: Blane HT, Leonard KE, eds. Psychological theories of drinking and alcoholism. New York: Guilford, 2007:181-226.

[7]   Martino SC, Collins RL, Ellickson PL, Schell TL, McCaffrey D. Socio-enviromental influences on adolescents' alcohol outcome expectancies: a prospective analysis. Addiction 2006;101(7):971-83.

[8]   Katz E, Fromme K, D' Amico E. Effects of outcome expectancies and personality on young adults' illicit drug use, heavy drinking, and risky sexual behaviour. Cognitive Ther Res 2000;24(1):1-22.

[9]   Baker T, Brandon T, Chassin, L. Motivational influences on cigarette smoking. Annu Rev Psychol 2004;55:463–491.

[10]   Chassin L, Presson CC, Sherman SJ, Edwards DA. Four pathways to young-adult smoking status: adolescent social-psychological antecedents in a Midwestern community sample. Health Psychol 1991;10(6):409-18.

[11]   Brandon TH, Baker TB. The smoking consequences questionnaire: The subjective expected utility of smoking in college students. Psychol Assessment 1991;3:484-91.

[12]   Chassin L, Presson CC, Young RD, Light R. Self-concepts of instructionalized adolescents: a framework for conceptualizing labeling effects. J Abnorm Psychol 1981;90(2):143-51.

[13]   Chassin L, Presson CC, Sherman SJ, Corty E, Olshavsky RW. Predicting the onset of cigarette smoking in adolescents: A longitudinal study. J Appl Soc Psychol 1984;14:224-43.

[14]   Hine DW, McKenzie-Richer A, Lewko J, Tilleczek K, Perreault L. A comparison of the mediational properties of four adolescent smoking expectancy measures. Psychol Addict Behav 2002;16(3):187-95.

[15]   Hine DW, Tilleczek K, Lewko J, McKenzie-Richer A, Perreault L. Measuring adolescent smoking expectancies by incorporating judgements about the expected time of occurrence of smoking outcomes. Psychol Addict Behav 2005;19(3):284-90.

[16]   Urbán R, Demetrovics Z. Smoking outcome expectancies: A multiple indicator and multiple cause (MIMIC) model. Addict Behav 2010;35:632-5.

[17]   Chung T, White H, Hippwell A, Stepp S, Loeder R. A parallel process model of the development of positive smoking expectancies and smoking behavior during early adolescence in Caucasian and African American girls. Addict Behav 2010;35:647-50.

[18]   Wahl S, Turner L, Mermelstein R, Flay B. Adolescent's smoking expectancies: Psychometric properties and prediction of behavior change. Nicotine Tob Res 2005;7(4):613-23.

[19]   Anderson CB, Pollak KI, Wetter DW. Relations between self-generated positive and smoking behavior: An exploratory study among adolescents. J Consult Clin Psychol 2002;70:998–1009.

[20]   Theodorakis Y, Hassandra M. Smoking and exercise, part II: Differences between exercisers and non-exercisers. Inquires Sport Phys Educ 2005;3(3):239-48.

[21]   European Commission and Statistical Office of the European Communities. Health statistics: Key data on health 2002: Data 1970–2001. Luxembourg: Office Official Publ Eur Commun, 2003.

[22]   Lewis-Esquerre J, Rodrigue J, Kahler C. Development and validation of an adolescent smoking consequences questionnaire. Nicotine Tobacco Res 2005;7(1):81-90.

[23]   Cohen L, McCarthy D, Brown S, Myers M. Negative affect combines with smoking outcome expectancies to predict smoking behavior over time. Psychol Addict Behav 2002;16(2):91-7.

[24]   Rosendahl KI, Galanti MR, Gilljam H, Ahlbom A. Knowledge about tobacco and subsequent use of cigarettes and smokeless tobacco among Swedish adolescents. J Adolesc Health 2005:37:224-8.

[25]   Kemm J. Health education: a case for resuscitation. Public Health 2003;117:106-11.

[26]   Gerrard M, Gibbons F, Benthin AC, Hessling RM. A longitudinal study of the reciprocal nature of risk behaviors and cognitions in adolescents: What you do shapes what you think, and vice versa. Health Psychol 1996;15(5):344-54.

In: Public Health Concern ISBN: 978-1-62948-424-2
Editors: J Merrick and A Tenenbaum © 2014 Nova Science Publishers, Inc.

Chapter VI

Sleep duration, smoking, alcohol and sleep quality

Bruce Kirkcaldy, PhD[*1] *and Timo Partonen, MD, PhD*[2]
[1]International Centre for the Study of Occupational and Mental Health,
Düsseldorf, Germany
[2]National Institute for Health and Welfare, Department of Mental Health
and Substance Abuse Services, Mood, Depression and Suicidal Behaviour
Unit, Helsinki, Finland

Abstract

1,276 adults completed a European internet survey detailing areas of the personal health, stress coping strategies and sleeping habits. Men reported sleeping shorter than women (6.76 vs. 6.99 hours per day) but complained "paradoxically" of having less quality sleep. Of the respondents, 35% were short sleepers (6 or less hours per day), 21% slept 8 hours per day and 4% were long sleepers (9 or more hours per day). The subjective reports of less satisfaction with the quality of sleep (sleeplessness) were predicted by the gender, the hours slept, the units of alcohol drunken, and the exercise taken (in hours per week). The age, the weekly hours of work, and the smoking behavior were not related to the quality of sleep. Gender differences in stress coping influenced the

* Correspondence: Bruce Kirkcaldy, PhD, International Centre for the Study of Occupational and Mental Health, Haydnstrasse 61, D-40595 Düsseldorf, Germany. E-mail: brucedavidkirkcaldy@yahoo.de.

sleeping behavior. These findings have both medical and psychological health care implications.

Sleep, that knits up the ravell'd sleeve of care.
The death of each day's life, sore labour's bath
Balm of hurt minds, great nature's second course.
Chief nourisher in life's feast.
"Macbeth" by Shakespeare (1564–1616)

Introduction

The health restoring and recuperating properties of sleep seem to have been well known throughout the centuries. Loss of quality of sleep and the adverse effects of sleep deprivation seem to be associated with inferior psychological and physical health. Current studies suggest that sleeping problems and sleep-related conditions are very common. Among some of the behavioral definitions of sleep disturbance are problems getting to or maintaining sleep, sleeping adequately but experiencing little in the way of rest and recuperation after waking, distress arising from recurring awakenings and detailed recall of frightening dreams involving personal threats, insomnia or hypersomnia complaints resulting from a reversal of the sleep-wake schedule normal for the individual's environment, or excessive daytime sleepiness and tiredness or bouts of falling asleep during the day (1).

One in 10 patients complains of sleep-related disorders when attending their general practitioner. Sateia (2) reviewed the literature of sleeping disorders indicating the incidence of occasional and intermittent insomnia to be around 35% for the general population, and the rates of chronic insomnia to be closer to 10% to 15%. European findings suggest the estimates of about 11% of the population. These numbers seem consistent across cultures. Ishigooka et al. (3) observed an average sleeping duration of 6.77 hours on weekdays, while approximately 20% revealed current sleeping problems, with 12% displaying insomnia that lasted over a month, in a large scale epidemiological outpatient study in Japan.

There appears to be an association between the disorders in sleeping behavior and the thought patterns. Borkovec and Roemer (4) found that it was the cognitive facet of anxiety, i.e. worrisome thoughts, rather than physiological symptoms of anxiety that was associated with insomnia. His suggestion for any effective treatment was to distract the individual away from intrusive thoughts to other sensations. People with insomnia tend to experience unwanted intrusive thoughts before sleep onset and tend to attribute difficulty falling asleep to their level of cognitive arousal (5). In addition, these same people spend much of their time concerned about insomnia and are likely to experience the same sort of intrusive thoughts about not being able to sleep and about hazardous consequences of sleep deprivation there might be. This in turn contributes to the general level of anxiety and discouragement. The intrusive thoughts about sleep are shaped by dysfunctional beliefs about sleep that are characteristic in people with insomnia.

OECDstatistics have demonstrated some cross-cultural differences in the sleeping behavior with the French sleeping the most (an average of 8.8 hours per day) followed by the US (8.6 hours per day). Norway, Sweden and Germany were the EU nations in which people slept the least. In Finland, females slept on average 7.62 hours in contrast to males with 7.39 hours on average. Short sleepers, as defined having 6 hours or less sleep per day, represented 15% of the sample, with more males being short sleepers (17% vs. 13% females), and about 1 in 7 (14%) was a long-sleeper (at least 9 hours per day). Overall, just over 1 in 5 persons reported experiencing sleeping disturbances. (http://www.oecd.org/statsportal/0,3352,en_2825_293564_1_1_1_1_1,00.html)

Hossain and Shapiro (6) claimed that difficulties falling asleep or daytime sleepiness commonly affects between 35% and 40% of the US adult population annually. They argue that the prevalence and burden in terms of personal and organizational health costs are significant, and sleeping ailments are often under-estimated and under-treated.

The prevalence rates of insomnia have been assessed in two Nordic countries, Norway and Finland, with a similar proportion of the population living at the same northerly latitudes. In one study, the one-month point prevalence of insomnia, as defined using the DSM-IV criteria, was 12% in a representative sample of the adult Norwegian population (7). Generally, sleep onset problems and daytime impairment were more common in winter than summer in this study, which was conducted over 12 months. Interestingly, the prevalence of sleep onset problems increased in southern Norway from summer to winter, while the opposite pattern was found in the northerly regions. Physical and mental health appeared to be the strongest predictors of insomnia in this study.

The second study provided similar results for the prevalence of insomnia as assessed with the use of DSM-IV criteria and the Sleep-EVAL system, which was 12%, in a representative sample of the adult Finnish population (8). An equal number of individuals reported global dissatisfaction with sleep. In general, sleep deterioration in summer or winter was linked to more complaints of sleep dissatisfaction. The prevalence of insomnia symptoms occurring at least three nights per week was 38%. Difficulty in initiating sleep was reported by 12%, difficulty in maintaining sleep by 32%, early morning awakenings by 11%, and non-restorative sleep by 8%. According to this study, insomnia is twice as prevalent in Finland as in other European countries, where insomnia was assessed by the same methods, but which are located closer to the equator.

Earlier, a questionnaire-based population survey of a Norwegian municipality (Tromsø) that lies north of the Arctic Circle included over 14,000 respondents. Among them, 42% of women and 30% of men said they were sometimes bothered by insomnia (9). Another survey in a general population north of the Arctic Circle indicated that 19% of women and 14% of men suffer mental distress during winter, and that among women insomnia in particular was most common when the length of day was at minimum (10). One-third of the women and one-fifth of the men experience problems with sleep, energy or mood in the high northern latitudes during winter. However, very few reported these problems in summer (11). In the very high north, sleeping problems lasting for at least two weeks are even more common, and this observation holds true in two dissimilar populations, Norwegians and Russians, living under similar Arctic conditions (12). Problems with falling asleep, not feeling rested the following morning, and waking up several times at night are the most frequent modes of sleeping problems in both populations. To a great extent, these problems may be the result of compromised adaptation, or inadequate acclimatization, after migration to the north.

In the re-analysis of all available data from the representative surveys carried out in Finland from 1972 to 2005, the main finding was that a decrease of self-reported sleep duration has taken place in Finland, especially among men of working age (13). The proportion of 7-hour sleepers has increased and, correspondingly, the proportion of 8-hour sleepers has decreased, but the extreme ends of the sleep duration distribution remained unchanged, yielding a general decrease in sleep duration by 18 minutes and an increase of sleep complaints among the employed middle-aged population in particular. The most critical factors for sleep duration appear to be the gender, the lack of energy, the insomnia, the marital status, the main occupation, the difficulties in getting sleep without a sleeping pill, and the leisure-time physical activity (14). Concerning the importance to the public health, e.g. due to decline in cognitive functioning (15), a detailed analysis of those having short or long sleep duration is needed for the evaluation of health hazards and a decision of preventive means.

The question is whether other countries, in this case Germany, have similar incidence rates of sleeping disorders. Another question is to what extent gender differences contribute. Certainly, as Steptoe et al (16) found in their analysis of the data on 18,000 students from 24 countries, marked differences emerged in sleeping duration across countries, so that concerning men for instance, Romanians slept the most (8.04 hours), then Spanish (8.02) and Greeks (7.86), and the nations in which people slept the least were Japan (6.09), Korea (6.89) and Taiwan (6.51). Statistics for the Germans among this young group (17-to-30-year old), revealed that on average women slept 7.60 hours and men 7.36 hours, and that short sleepers exhibited poorer self-rated health among both men and women, whereas long sleep duration was not related to the self-reported health status.

Our main research questions herein are: (a) Do women report sleeping more hours than men? (b) Are there gender differences in the quality of sleep, i.e. the subjective reports of insomnia? (c) What are the common and gender-specific coping strategies that associate with sleeping disorders, i.e. insomnia and/or extreme sleep durations? (d) What are the socio-demographic predictors of sleeplessness? (e) Do short and/or long sleepers differ in their physical and psychological health? (f) Is the relationship between distress and insomnia moderated by any coping strategy?

Our study

In a 3-month period, an internet-based exploratory research study was carried out collating data from a new psychometric instrument. Recent studies suggest that comparative analyses about Internet-operated survey questionnaires reveal findings that can be generalized across presentation formats, are not adversely affected by non-serious or repeat responders, and produce results that are consistent with traditional methods of data collation (17). Internet certainly presents empirical researchers with tremendous opportunities for collecting large data bases on topics of contemporary interest (18). In addition to details of sleeping habits such as sleep duration (in hours) and the quality of sleep, persons were required to assess, with a 1-to-5 Likert scale, the degree of endorsement of items that were associated with stress coping strategies (28 items). The alpha-coefficient for the total coping scale was 0.75. A further set of 27 items constituted the stress scale, with the alpha coefficient for total score being 0.91. The two health outcome scales, mental (ill) health and psychological (ill) health, had the alpha coefficients of 0.85 and 0.93 respectively and comprised of 16 items each. Moreover, questions were asked regarding socio-demographic data, including the age, the gender, the educational status and the marital status. Using anonymous administration of survey questionnaires might have allowed us to have better estimates of the actual compulsive thinking.

One of the difficulties inherent in such data collection involves the issue of the quality of data. In order to ensure good quality data sets in this study, the data were double-screened, checked through visually and through descriptive analyses filtering out multiple responders and removing those respondents with the incomplete data or those who exhibited obvious outliers or response sets. Individuals were provided with a special coded number after the completion of the survey enabling them to obtain their detailed profile scores if requested. During a period of 6 months, there were 1,275 persons who had completed the entire survey questionnaire in complete. Their average age was 35.54 years (SD=9.01) and average number of children, including those persons having no children, was 0.63 (SD=0.99). The distribution of the demographic data and the size of the sample suggest that this data set yields valid and reliable results.

Findings

Overall, the average number of hours slept per day was 6.83 (SD=1.12) for 1204 participants. Women reported sleeping longer (6.76, SD=1.13) than men (6.99, SD=1.12), yielding a difference that was significant ($F_{(1,1202)}=11.65$, $P<0.001$).

8.2% of the sample (n=99) reporting sleeping five or less hours per day on average (2% reported to sleep for 4 or less hours). More than a quarter (26.3%, n=317) reported sleeping 6 hours per day, and almost two-thirds (61.6%) between 7 and 8 hours (40.3% slept 7 hours per day). Only 3.9% (n=47) reported sleeping 9 or more hours per day.

Table 1. Predictors of insomnia

Model	$R^2=0.27$, adjusted R2=0.07	F(5,818)=13.25	P<0.001***
	beta	t	P
Hours slept per day	-0.19	-5.69	0.001
Units of alcohol consumed	+0.16	4.67	0.001
Gender (women)	+0.13	3.71	0.001
Exercise per week in hours	-0.07	-2.08	0.05
Body-mass index	+0.07	2.02	0.05

Examining the proportions of those reporting problems that were associated with insomnia, over a third (35.6%) had no problems with sleeping. There were occasional "once or twice per year" (33.3%), monthly (13.2%), weekly (11.5%) and almost daily "very regularly" (6.4%) complaints. Overall, the proportion of persons suffering from sleeping problems (daily or weekly) was 17.9% in total. The incidence of these regular bouts of sleeplessness were higher among women (23.2%) than men (15.2%), yielding a difference that was significant (chi-squared (4)=14.04, P<0.01). Hours slept and insomnia were correlated albeit the magnitude of the correlation was moderate (r=-0.22, P<0.001). The predictors of insomnia are presented in Table 1.

Table 2 reveals the predictors of sleep duration, after using multiple linear regression analysis. Two coping strategies that were common to men and women associated with sleep duration: "talking with others" and "smoking more". There were also gender-specific determinants, as e.g. self-blaming, problem-focused, and procrastination (putting off decisions) did relate to a shorter number of hours slept among women.

Table 3 reveals that the coping styles that determined the sleep duration were not the same as those predicting insomnia. Women prefer to use coping strategies that men appear not to. Worrying was a determinant of insomnia for men, but not for women.

Next, we wanted to see whether men and women differed in their frequency of using coping methods. A linear discriminant analysis revealed that there were significant gender differences in the coping profiles (F(12,1273)=23.25, P<0.001; R=0.43, Wilks lambda=0.82, chi-squared(12)=253.01, P<0.001). The difference (the significant beta loadings are given in parentheses) was the greatest for talking with others (0.43), a change in eating habits (0.32), seeking support (0.29), prioritizing (0.29), worrying (0.27) and day-dreaming (0.26) that women used more often as a coping strategy. In contrast, men were more likely to use solution focusing (-

0.26), a drink of alcohol (-0.25), taking breaks (-0.22) and wishful thinking (-0.21) as a coping strategy.

Table 2. Coping determinants of sleep duration

Model	$R^2=0.27$, adjusted $R^2=0.07$	$F(7,791)=8.96$	$P<0.001$***
Men	Beta	t	P
Positive self-talk	0.11	3.14	0.002**
Increased smoking	-0.11	-2.96	0.003**
Talk with others	0.12	3.40	0.001***
Avoid situations/people	0.13	3.71	0.001***
Wishful thinking	-0.11	-2.88	0.004**
Taking breaks	0.09	2.45	0.02*
Seeking social support	-0.08	-2.35	0.02*
Model	$R^2=0.29$, adjusted $R^2=0.07$	$F(5,398)=7.13$	$P<0.001$***
Women	Beta	t	P
Increased smoking	-0.15	-3.04	0.003**
Talk with others	0.16	3.18	0.002**
Self-blame	-0.11	-2.29	0.03*
Solution (focused)	-0.15	-2.95	0.003**
Putting off decisions	-0.12	-2.36	0.02*

Table 3. Coping determinants of insomnia

Model	$R^2=0.38$, adjusted $R^2=0.14$	$F(5,834)=27.65$	$P<0.001$***
Men	Beta	t	P
Withdrawal	0.18	5.19	0.001
Worrying	0.15	4.46	0.001
Drinking	0.11	3.42	0.001
Day-dreaming	0.12	3.38	0.001
Prioritizing	-0.08	-2.52	0.05
Model	$R^2=0.46$, adjusted $R^2=0.20$	$F(7,426)=16.37$	$P<0.001$***
Women	Beta	t	P
Self-blame	0.23	4.75	0.001***
Taking drugs/medicine	0.09	1.78	0.10
Smoking more	0.12	2.79	0.01**
Building a solution	-0.09	-1.95	0.06
Taking breaks	-0.10	-2.32	0.03*
Withdrawal	0.11	2.24	0.03*
Drinking alcohol	0.10	2.22	0.03*

Stepwise linear regression analysis revealed that five of the variables emerged as statistically significant (non-significant were the age, weekly hours in work and smoking habits). In other words, the more hours slept per day, the less insomnia. Similarly, the insomnia experienced was lower when the regular exercise was more frequent. In contrast, the insomnia was higher when the body-mass index was greater, and its (antithetical) effect was equal to that of regular exercise. Finally, the units of alcohol were associated with insomnia.

A correlation analysis (n=1204) revealed that, in contrast to insomnia, sleep duration was significantly correlated with the number of cigarettes smoked per day (r=-0.12, P<0.001), age (r=-0.10, P<0.001) and weekly hours in work (r=-0.08, P<0.01).

Finally, a series of ANOVAs was computed separately for sleep duration and insomnia. Low and high scoring stress and coping groups were generated using a cut-off of 0.5 of the standard deviation from the mean. The low and high stress groups were subsequently different in the terms of overall stress scores (t(432)=-28.91, P<0.001) as were the coping (low and high) groups in the terms of overall coping scores (t(433)=-27.04, P<0.001). The potential confounding effect of the covariate gender differences was controlled for. For both the quantity and quality of sleep, the main effect due to distress was statistically significant, but none of the main effects attributable to coping were significant. Moreover, there was no significant stress × coping interaction (Table 4). For example, there was no difference in sleep duration or insomnia between the low and high coping groups, but the more stressed (S+) persons slept shorter and complained more insomnia than the low stress (S−) persons. There was no indication that persons with the more frequent implementation of coping strategies could moderate the potentially deleterious effects of perceived stress.

Table 4. Duration of sleep (hours slept) and quality of sleep (insomnia) for the low and the high scoring groups by distress and coping

	S-C-	S-C+	S+C-	S+C+	Stress	Coping	Stress × Coping
Hours slept	7.16 (1.05)	7.16 (1.13)	6.72 (1.17)	6.84 (1.13)	11.25***	0.27	0.27
Insomnia	1.96 (1.02)	2.18 (1.18)	2.71 (1.45)	2.75 (1.39)	28.49***	1.13	0.55

Next, we moved onto the outcome variables: mental and physical health. Prior to do so, we generated five groups of sleepers (based on sleep duration). We collapsed the data for persons with 5 or less hours of sleep (n=99, 8.2%),

26.3 percent reported 6 hours of sleep (n=317), 40.3% had 7 hours of sleep (n=485), 21.3% had 8 hours of sleep (n=256), and 3.9% were long sleepers (9 or more hours per day, n=47). Separate analyses were conducted using ANOVAs, with the gender and the sleep categories as the independent variables, for each outcome health variable.

Figure 1 demonstrates the effect of sleep duration on physical (ill) health for men and women. The gender effect was significant ($F(1,1193)=27.93$, $P<0.001$), as was the main effect of sleep duration ($F(4,1193)=17.15$, $P<0.001$), but there was no significant interaction term ($F(4,1193)=0.62$, $P>0.05$). The plot reveals that both men and women display an improvement in physical health with the increase in sleep time up to 8 hours, and thereafter longer sleep durations lead to a slight increase in ill-health.

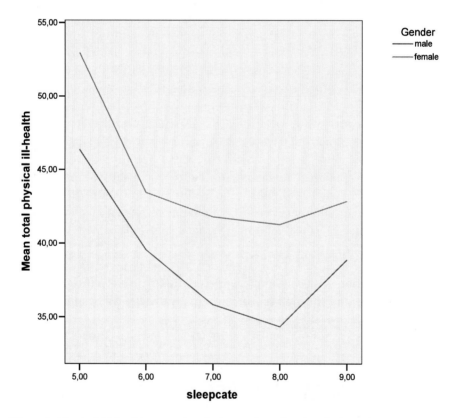

Figure 1. Physical ill-health scores in relation to sleep duration (in hours).

With respect to psychological (ill) health, a clear gender difference is observed, with women reporting poorer health than men ($F(1,1193)=23.41$, $P<0.001$). In addition, there was a significant main effect of sleep duration, indicating that especially for the short sleepers (5 or less hours) psychological health is the worst ($F(4,1193)=8.40$, $P<0.001$). A tendency for the gender × sleep duration interaction ($F(4,1193)=2.06$, $P<0.10$) to be significant reflects the gender-specific difference in sleep duration, with women's psychological health leveling off after 6 or more hours, and with men's psychological health continuing to improve up to 8 hours (men's need for a longer sleep) and those who were long sleepers (9 or more hours) having a slight increase in psychological ill-health (see Figure 2).

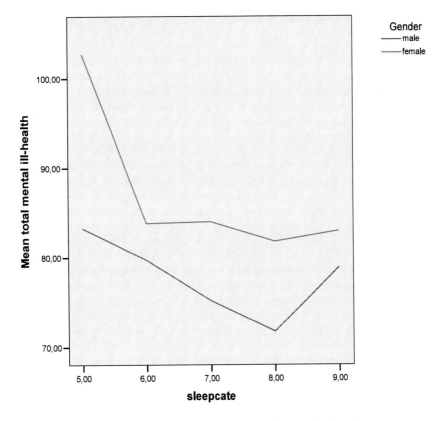

Figure 2. Mental ill-health scores in relation to sleep duration (in hours).

Discussion

Our key finding here was that shorter sleep durations (reduced quantities of sleep) were associated with not only poorer physical health but also poorer psychological health. These findings bear implications for social and medical policy makers in their work for the prevention of sleeping problems and the subsequent health hazards.

The unhealthy lifestyles altogether such as obesity, lack of exercise, excessive alcohol use and smoking have influence on the quantity and quality of sleep. Among the reasons that exercise appears to promote healthy sleeping habits is the findings that insomniacs appear to be more sedentary than those persons who report sleeping well. It is likely that a deficiency in physical activity can contribute to insomnia by "inhibiting the daily rise and fall of the body temperature rhythm. As a result, many get caught in a vicious circle of insomnia, reduced energy, decreased physical activity, and worsened insomnia" (19). Jacobs (19) further argued that because of the temperature increasing properties of physical activity, followed by a compensatory reduction after a couple of hours rest, the latter facilitates sleep. This is particularly pronounced when exercise with within 3-6 hours prior to going to bed. The finding in our study that not only exercise but body mass index was associated with quality of sleep is underlined by the elevated BMI, presumably due to more sedentary life of non-exercisers.

Penn and Zalaesne (20) discuss the surprising implication in the study linking between obesity (high body mass index) and sleeplessness. Overweight can result in sleeping problems, including blocked airways which constrain breathing. Inadequate sleep in addition triggers hormones which increase the drive for hunger and appetite, implies that sleeping too little will increase the likelihood of getting fat. However, a key to weight gain and weight management seems to be eating in the evening or before sleep rather than the duration of sleep, as caloric intake after 8 p.m. increases the risk of obesity independent of sleep timing and duration (21), and as changes in the peripheral levels of the hormones driving for hunger and appetite are not enough to explain the overeating after insufficient sleep (22). Curiously, women managed to maintain weight during adequate sleep to meet their sleep need, while insufficient sleep reduced the dietary restraint and led finally to weight gain in women as well (22).

Therefore, there needs to be a centrally guided drive to increase eating in order to meet the body's increased energy demands. This implies that the master circadian clock has a key role here, and that it is the prolonged sleep restriction with concurrent circadian misalignment that alters the ratio of energy storage to energy expenditure in favour of the excessive energy intake and the subsequent weight gain. In fact, circadian misalignment, whether due to a phase advance or a phase delay, leads to sleep fragmentation, to an increase in time spent awake after sleep onset, and to a shortened rapid-eye-movement sleep latency (23). These are typical findings of sleep in the depressed individuals.

Sleeping for only 6 hours per night will increase the risk of developing obesity by 23%, and further reduction to about 4 hours sleep elevates the risk by 73% as reported from the "2005 Sleep in American Poll" (www.sleepsolutions.com/phys/education/NSF_2005_Sleep_in_America_Poll _Results.htm). Our finding of the body-mass index being a significant independent predictor of insomnia agrees with this link between sleep duration and weight.

Determinants of sleeping disorders and disruptions include the effects of alcohol and nicotine. These predictors of sleeping disorders can be generalized across nations. Alcohol has been referred to by Jacobs (19) as the "night cap that causes insomnia". In cases where alcohol may relax people sufficiently that they may fall to sleep more easily, problems arise from the fragile and fragmented sleep which disrupts quality of sleep, restorative sleep. Consumption of alcohol initially has sedating properties and thus the immediate short term subjective benefits may be experienced in the feeling of drowsiness and fatigue, but overall alcohol upsets sleep by disrupting patterns such as the sequence of duration of states of sleep (slow wave sleep and rapid eye movement) and adversely affects sleep latency (time required to fall asleep) and the duration of sleeping time (NIAAA Alcohol Alert No. 41 July 1998 on www.niaaa.nih.org). Jabobs (19) further mentions how alcohol disrupts sleep because of the time required for its metabolism and the subsequent withdrawal symptoms.

Concerning nicotine, no relationship was found between the number of cigarettes smoked daily and the quality and duration of sleep. The lack of relationship may be because the use of smoking as a method of coping with stress may be gender specific, that is, we found men who prefer this style of coping were likely to sleep for shorter duration. For women, smoking more was associated with a higher degree of insomnia but did not relate with sleep duration. In a Japanese study (24), analysis of the effect of the intensity of smoking on sleeping latency and quality demonstrated that the global sleep quality index was much higher (inferior sleep quality) in male ex-smokers vs. non-smokers and among female current smokers vs. non-smokers. Men who presently smoked (21 to 40 cigarettes per day) and ex-smokers displayed a significant increase in terms of poor sleep. For women, only current smokers exhibited a high incidence of inferior sleep. Smokers have more problems going to sleep, staying asleep and with daytime sleepiness (25). Our study suggests that advice in nutritional and physical health parameters in adequate treatment of sleeping disorders may have a major role. In particular, as it has been demonstrated by others that not only insomnia (26), but also smoking (27) leads to weigh gain.

Both women and men having longer sleep durations reported of talking with others as a coping strategy. Interestingly, in contrast to women, men had a range of strategies, including positive self-talk, taking a break and avoiding situations or people, that bear elements of cognitive and behavioral actions, similar to psychological interventions that have a focus on cognitive schemas and behavioral conditioning. Those having shorter sleep durations use coping strategies that can be negative or positive in control for insomnia. The former include the increased smoking and self-blame among women, and the increased smoking and wishful thinking among men. Women used focused solution and putting off decisions and men seek social support as positive means for coping with shorter sleep durations.

Shorter sleep durations often precede shortened sleep durations, i.e. insomnia, and we found here that withdrawal and drinking alcohol had a negative effect on sleep. In addition, women with insomnia had self-blame, increased smoking and took sleeping pills. Men with insomnia reported of worrying and day-dreaming. Proactive means for coping with insomnia included taking a break and focused solution among women, whereas men used prioritization.

Our analysis of the low and high scoring stress and coping groups indicated that persons with the more frequent implementation of coping strategies could not moderate the perceived distress, and thereby suggests that practical solutions for stress reduction, therapeutic processes and treatment of patients in general are likely to have a major role in the prevention of sleeping problems. However, physical conditions and mental disorders such as depression in particular are highly co-morbid with sleeping problems that makes it a challenge to develop such solutions. These solutions may be provided by the preference to options that include cognitive, behavioral and emotive elements of intervention, e.g. rational-emotive therapy or cognitive-behavioral therapy of insomnia. For example, insomnia may have a corrective homeostasis function (28, p. 255) that is a kind of paradoxical intention. A symptom, e.g. insomnia, is welcomed, a kind of cognitive reframing, in which the physiology seeks balance and distracts away from the frightening and potentially devastating thoughts associated with being "sleep disturbed". Future studies may focus on "content" of worry, identification of cognitions associated with sleeping problems. We used one scaling of worrying as a coping technique without distinguishing content of thoughts.

There are limitations to our study. One limitation is that we cannot deduce any direction of causality, however, because of the cross-sectional design of our study. On the one hand, poorer health may easily result in the increase in sleeping disorders. On the other hand, it may well be that shorter sleep durations induce fatigue and exhaustion throughout the day and indeed compromise the recuperating value of sleep-related physiological processes. We have witnessed how other correlates of insomnia may have confounded the relationship such as physical activity, alcohol consumption and gender. Our findings here are consistent with those of Steptoe et al. (16) in that the adverse effects of sleep duration may not hold for long sleepers, although gender moderated the relationship and the link was further influenced pending on whether one considered physical or mental health as the outcome variable. Moreover, our finding agrees with those reporting women to sleep longer than men, by 15 minutes on average. In the younger group of students, Steptoe et al. (16) reported a difference of 12 minutes, 7.6 vs. 7.4 hours, and the longer duration herein could be attributable to age, our study having the mean of 35 years in contrast to the mid-twenties for the European Behavior Survey. About a third (36%) of our sample slept less than 7 hours per day. Insomnia and sleep duration correlated but the magnitude was moderate.

Another limitation is reliance on the subjective reports of sleeping behavior. How reliable a self-report is? For future studies, we will definitely need a more comprehensive list of variables, e.g. the estimated sleep onset, the number of interruptions per night, working hours, leisure pursuits, etc., that are known to be associated with sleep duration and sleeping behavior to be able to fully adjust our analysis for confounding factors.

In conclusion, shorter sleep durations were associated with not only poorer physical health, but also poorer psychological health. Persons having shorter or shortened sleep durations tend not to use means that have control of their distress. Recent research (29) suggests that rather than psychological disorders resulting in sleeping disorders, it may be that sleep deprivation is causal on behavioral and emotional disorders (presumably including addictive and compulsive behaviors as well as bipolar disorders, depression and PSTD), so that "an amplified hyperlimbic reaction by the human amygdala was observed in response to negative emotional stimuli... this altered magnitude of limbic activity is associated with a loss of functional connectivity with the mPFC (*medial prefrontal cortex*) in the sleep deprivation condition implying a failure of top-down inhibition by the prefrontal lobe .. (and) a night of sleep may "reset" the correct affective brain reactivity to next-day emotional challenges by maintaining functional integrity of this mPFC–amygdala circuit and thus govern appropriate behavioral repertoires (e.g., optimal social judgments and rational decisions)." (p.186). Hence clinicians may consider "targeted sleep restoration in anxiety may ameliorate excessive anticipatory responding and associated clinical symptomatology" (30).

Supporting information table. The basic description of the sample.

	%	n
Gender		
Male	65.8%	843
Female	34.2%	433
Age		
<20 years	1.3%	16
21-30	34.0%	434
31-40	37.3%	476
41-50	20.4%	260
51-60	6.4%	82
>60 years	0.6%	8
Educational status		
Secondary	23.5%	298
GCE A level (12th grade)	22.3%	283
First degree	23.7%	300
Higher degree	24.0%	304
Other	6.5%	82
Marital status		
Single	30.6%	388
Married or cohabiting	61.7%	782
Divorced	4.3%	55
Separated	2.8%	36
Widowed	0.5%	6

References

[1]   Jongsma AE, Peterson LM. The complete adult psychotherapy treatment planner. 2nd ed. New York: John Wiley, 1999.

[2]   Sateia MJ. Description of insomnia. In: Kushida CA, editor. Handbook of sleeping disorders. 2nd ed. New York: Informa Healthcare USA, 2008:3-15.

[3]   Ishigooka J, Suzuki M, Isawa S, Muraoka H, Murasaki M, Okawa M. Epidemiological study on sleep habits and insomnia of new outpatients visiting general hospitals in Japan. Psychiatry Clin Neurosci 1999;53:515-22.

[4]   Borkovec TD, Roemer L. Perceived functions of worry among generalized anxiety disorder subjects: distraction from more emotionally distressing topics? J Behav Ther Exp Psychiatry 1995;26:25-30.

[5]   Clark DA, editor. Intrusive thoughts in clinical disorders: theory, research, and treatment. New York: The Guilford Press, 2004.

[6]   Hossain JL, Shapiro CM. The prevalence, cost implications, and management of sleep disorders: an overview. Sleep Breath 2002;6:85-102.

[7]   Pallesen S, Nordhus IH, Nielsen GH, Havik OE, Kvale G, Johnsen BH, Skjøtskift S. Prevalence of insomnia in the adult Norwegian population. Sleep 2001;24:771-9.

[8]   Ohayon MM, Partinen M. Insomnia and global sleep dissatisfaction in Finland. J Sleep Res 2002;11:339-46.

[9]   Husby R, Lingjaerde O. Prevalence of reported sleeplessness in northern Norway in relation to sex, age and season. Acta Psychiatr Scand 1990;81:542-7.

[10]   Hansen V, Jacobsen BK, Husby R. Mental distress during winter: an epidemiologic study of 7759 adults north of Arctic Circle. Acta Psychiatr Scand 1991;84:137-41.

[11]   Hansen V, Lund E, Smith-Sivertsen T. Self-reported mental distress under the shifting daylight in the high north. Psychol Med 1998;28:447-52.

[12]   Nilssen O, Lipton R, Brenn T, Höyer G, Boiko E, Tkatchev A. Sleeping problems at 78 degrees north: the Svalbard Study. Acta Psychiatr Scand 1997;95:44-8.

[13]   Kronholm E, Partonen T, Laatikainen T, Peltonen M, Härmä M, Hublin C, Kaprio J, Aro AR, Partinen M, Fogelholm M, Valve R, Vahtera J, Oksanen T, Kivimäki M, Koskenvuo M, Sutela H. Trends in self-reported sleep duration and insomnia-related symptoms in Finland from 1972 to 2005: a comparative review and re-analysis of Finnish population samples. J Sleep Res 2008;17:54-62.

[14]   Kronholm E, Härmä M, Hublin C, Aro AR, Partonen T. Self-reported sleep duration in Finnish general population. J Sleep Res 2006;15:276-90.

[15]   Kronholm E, Sallinen M, Suutama T, Sulkava R, Era P, Partonen T. Self-reported sleep duration and cognitive functioning in the general population. J Sleep Res 2009;18:436-46.

[16]   Steptoe A, Peacey V, Wardle J. Sleep duration and health in young adults. Arch Intern Med 2006;166:1689-92.

[17]   Gosling SD, Vazire S, Srivastava S, John OP. Should we trust web-based studies? A comparative analysis of six preconceptions about Internet questionnaires. Am Psychol 2004;59:93-104.

[18]   Kraut RM, Olson J, Banaji M, Bruckman A, Cohen J, Couper M. Psychological research online: report of Board of Scientific Affairs' Advisory Group on the conduct of research on the Internet. Am Psychol 2004;59:106-17.

[19]   Jacobs GD. Say goodnight to insomnia. New York: Henry Holt and Co., 1999:107-8.

[20]   Penn J, Zalesne EK. Micro trends: surprising tales of the way we live today. New York: Penguin, 2007.

[21]   Baron KG, Reid KJ, Kern AS, Zee PC. Role of sleep timing in caloric intake and BMI. Obesity (Silver Spring) 2011;19:1374-81.

[22]   Markwald RR, Melanson EL, Smith MR, Higgins J, Perreault L, Eckel RH, Wright KP Jr. Impact of insufficient sleep on total daily energy expenditure, food intake, and weight gain. Proc Natl Acad Sci U S A 2013;110:5695-700.

[23]   Gonnissen HK, Mazuy C, Rutters F, Martens EA, Adam TC, Westerterp-Plantenga MS. Sleep architecture when sleeping at an unusual circadian time and associations with insulin sensitivity. PLoS ONE 2013;8:e72877.

[24]   Hu L, Sekine M, Gaina M, Kagamimori S. Association between sleep quality and smoking in Japanese civil servants. Sleep Biol Rhythms 2007;5:196-203.

[25]   Phillips BA, Danner FJ. Cigarette smoking and sleep disturbance. Arch Intern Med 1995;115:734-7.

[26]   Spiegel K, Tasali E, Leproult R, Van Cauter E. Effects of poor and short sleep on glucose metabolism and obesity risk. Nat Rev Endocrinol 2009;5:253-61.

[27]   Saarni SE, Pietiläinen K, Kantonen S, Rissanen A, Kaprio J. Association of smoking in adolescence with abdominal obesity in adulthood: a follow-up study of 5 birth cohorts of Finnish twins. Am J Public Health 2009;99:348-54.

[28]   Bernard ME, editor. Using rational-emotive therapy effectively: a practitioner's guide. New York: Plenum Press, 1991.

[29]   Walker, M. The role of sleep in cognition and emotion. The Year of Cognitive Neurosciences. Ann. New York Academy of Sciences, 2006, 1156, 168-197.

[30]   Goldstein A, Greer S, Saletin JM, Harvey AG, Nitschke JB, Walker MP. Tired and apprehensive. Anxiety amplifies the input of sleep loss on aversive brain anticipation. Journal of Neurosciences, 2013; 33, 28, 10607-10615.

In: Public Health Concern   ISBN: 978-1-62948-424-2
Editors: J Merrick and A Tenenbaum   © 2014 Nova Science Publishers, Inc.

Chapter VII

Cigarette smoking among Nigerian adolescent school boys

Alphonsus N Onyiriuka, MD and Rita C Onyiriuka BSc*
Department of Child Health, University of Benin Teaching Hospital,
Benin City, and Department of Economics and Statistics,
Faculty of Social Sciences, University of Benin, Benin City, Nigeria

Although recent trends indicate a rising cigarette-smoking prevalence among adolescents in African countries (Nigeria inclusive), data on characteristics and attitudes of these African adolescent smokers are scarce. In this chapter an anonymous self-administered questionnaire was used in obtaining data on cigarette smoking from 1,060 male students in two boys' secondary schools. Prevalence of cigarette smoking was 12.0%, 22.4% and 17.1% among first-year, second-year and third-year boys respectively, with an overall prevalence of 16.5%. The mean age at which smoking began was 16.9 ± 1.2 years (95% confidence interval, CI $= 16.7 - 17.1$). A significantly higher risk of smoking was associated with having parents, elder siblings or best friends who smoked. Tolerant attitude of parents and best friends encouraged smoking among students. The two leading reasons for taking up smoking were peer pressure effect (36.5%) and the need to mix freely with peers in social situations (22.9%). All the smokers were "light smokers" (smoke < 10 sticks of

* Correspondence: Alphonsus N Onyiriuka, MD, Department of Child Health, University of Benin Teaching Hospital, PMB 1111, Benin City, Nigeria. E-mail: alpndiony@yahoo.com or didiruka@gmail.com.

cigarette per day) but 3.4% smoked their first cigarette within 30 minutes of waking up. Among the smokers, 80.6% admitted smoking was harmful and 60% said they would like to quit smoking but lacked the will power and were uncertain how to quit. Considering that the prevalence of smoking among adolescents tended to increase with age, school-based smoking control intervention should start in the primary/junior secondary schools and it must also address family influences on adolescent smoking. Empowering adolescents with skills to resist peer pressure is advocated.

Introduction

Cigarette-smoking often begins during the period of adolescence and its prevalence is increasing. It constitutes a public health problem, not only because of the preventable morbidity and mortality it causes, but also, because of its socioeconomic implications (1,2). The prevalence rate of cigarette smoking is declining in developed countries (3), but rising in developing countries (4). Despite this trend, there is paucity of information on cigarette smoking among adolescents in most developing countries (5). Cigarette smoking commonly begins during adolescence when several factors such as peer pressure, family influences, social class, curiosity, defiance of social norms and the desire to imitate others influence a youth to start smoking and maintain the habit (6,7).

Adverse effects of cigarette smoking may occur during adolescence. For instance, increased prevalence of chronic cough, sputum production, wheezing, malodorous breath and yellowish discoloration of the teeth have been documented (8). Cigarette smoke is known to induce hepatic smooth endoplasmic reticulum enzymes which may influence metabolism of some drugs such as phenacetin, theophylline and imipramine and also endogenously produced hormones (8). Studies have shown that adolescent cigarette smokers are at increased risk of other forms of drug misuse (9-11). Inhalation of cigarette smoke by adolescents predispose to early pharmacological dependence on cigarette (10). It is known that adolescent smokers may become nicotine dependent while smoking fewer number of cigarette sticks per day than are necessary for adult smokers (8). More than 90% of adolescent smokers become adult smokers (8). It has been shown that signs of nicotine dependence include smoking more than 10 cigarettes per day or feeling a craving to smoke their first cigarette within 30 minutes after waking up (12,13).

The reported prevalence of cigarette smoking among male secondary school students varied from country to country. For instance, the following prevalence rates have been reported: Republic of Yemen 21.9% (14); Saudi Arabia 20.0% (15); Syria 15.9% (16); Portugal 21.5% (9); and Bahrain 21.0% (17).

In Nigeria, there is paucity of information on the prevalence of cigarette smoking, smoking behaviours and attitudes toward smoking among our adolescent secondary school students (18). Ibe and Ele (19), in Nigeria, studied only female secondary school students despite the reported higher prevalence among boys (5,9,20). In this regard, data from Nigeria on cigarette smoking among male secondary school students will be intriguing given the fact that our social, cultural and economic patterns differ significantly from those of many non-African countries and even, some African countries. The purpose of the present study was to determine the prevalence of cigarette smoking among adolescent senior secondary school boys in Oredo LGA and assess the characteristics and attitudes of the smokers. The data obtained will provide a base-line of information needed for planning policies and strategies for school cigarette-smoking control programs.

Our study

The study was a cross-sectional survey conducted in the two public senior secondary schools for boys in Oredo Local Government Area (LGA), Edo State. According to Edo State Ministry of Education Statistics (21), there are nine public senior secondary schools in the LGA comprising four for only girls, three co-educational (boys and girls mixed) and the remaining two for only boys. In each of the two boys' senior secondary school, all classes were recruited according to schedule convenience of each class to minimize disruption of the academic activity of the class. All students in each of the classes were asked to participate. Consent for the study was obtained from the school authorities.

Data for the study was collected between January and February 2009, using anonymous self-administered questionnaire based on WHO guidelines (22) for conduct of survey on smoking prevalence and behaviour with some additional questions. The students completed the questionnaires in their classrooms and the completed questionnaires were collected on the spot and were analyzed at the end of the study period.

In the survey design, we emphasized to the students that their participation was voluntary and the questionnaire was anonymous in order to protect the students' privacy and minimize underreporting. In this survey, information was collected on socio-demographic characteristics (e.g., age, class), smoking behaviour (e.g., smoking status, age at which smoking started, number of cigarettes smoked per day, reasons for smoking, place of smoking, place of procurement of cigarette, whether or not younger siblings go to purchase cigarette for them and the time they smoke their first cigarette on waking up in the morning and whether or not they smoke in the presence of their parents). The attitude of smokers toward the harmful nature of smoking and quitting of smoking was assessed. We also documented the disapproving attitude or otherwise of the smokers' close contact toward smoking. The socio-economic classification of each student was determined using a scoring system suggested by Ogunlesi et al (23), which combined the highest educational attainment, occupation and income of the parents. The scores for both parents were used to assign their children to socio-economic classes (I to V) with social classes I and II representing high social class, social class III representing middle social class while social classes IV and V representing low social class.The students' individual academic performance was assessed by classifying them into three levels. "never failed" "failed once" and "failed more than once". In this study, each failure refers to having to repeat one year at school. The smokers were also assessed for signs of nicotine dependence by identifying those who smoked 10 or more cigarettes per day or feel a craving to smoke their first cigarette within 30 minutes after waking up in the morning. Those with any of these two symptoms were regarded as potentially nicotine dependent (12,13), requiring initiation of treatment.

Data management and analysis were performed using the SPSS statistical program. Chi-square test was used for assessing the significance of differences between smoking status and study variables among respondents. A p-value <0.05 was considered significant. Odd ratio (OR) and 95% confidence intervals (CI) were calculated.

Smokers were those who smoked at the time of the study, whether daily, weekly or less than weekly. Ex-smokers were those who have smoked before but were not smoking at the time of the study. "Non smokers" were those who have never smoked and ex-smokers combined (22). Those who smoked less than cigarettes per day were classified as "light smokers" while those who smoked 10 or more cigarettes per day were classified as "heavy smokers".

Findings

At the time of the survey, a total of 1,073 boys (974 in school A and 99 in school B) were attending the two boys' senior secondary schools in the Local Government Area (LGA). Of this number, 6 students declined to participate in school A while 4 students declined to participate B, yielding a response rate of 99.4% in school A and 96.0% in school B. Overall response rate was 99.1%. The questionnaires of 3 students were excluded from the analysis because they were incompletely filled, leaving a total of 1,060 questionnaires whose data are presented. Students in both schools had similar socio-demographic characteristics.

Of the 1,060 respondents, 175 (16.5%) were smokers while the remaining 885 (83.5%) were non-smokers, comprising 79 (8.9%) ex-smokers and 806 (91.1%) never-smokers. As shown in Table 1, prevalence was higher among students in their third year (SS 3) and older (aged between 17 and 21 years) and from families in the middle social class. Both smokers and non-smokers had similar economic and demographic profiles. Out of 175 smokers, 154 (88.0%) purchased their cigarette from either a nearby shop or hawker. Table 2 showed that peer group pressure and the need to mix freely with peers in social situations, together accounted for 59.4% of the reasons for smoking by respondents.

Table 1. Socio-demographic variables and prevalence
of cigarette-smoking

Socio demographic variable	No of respondents (%)	No of smokers	Prevalence in percent
School year			
First year (SS1)	366 (34.5)	37	10.1
Second year (SS2)	359 (33.9)	53	14.8
Third year (SS3)	335 (31.6)	85	25.4
Total	1060 (100)	175	16.5
Age group (years)			
13 – 15	170 (16.0)	17	10.0
16 – 17	715 (67.5)	111	15.5
18 –21	175 (16.5)	117	66.9
Total	1060 (100)	175	16.5
Social class of family*			
High social class	335 (61.6)	33	9.9
Middle social class	412 (38.9)	74	18.0
Low social class	313 (29.5)	47	15.0
Total	1060 (100)	164	15.5

* Eleven (6.3%) of the smokers have lost one of their parents, as a result were excluded from the social class determination.

The mean age at which smoking commenced was 16.9 ± 1.2 years (95% confidence interval, CI = 16.7 − 17.1). Table 2 showed that all the smokers were "light smokers" (smoked less than 10 sticks of cigarette per day) but 3.4% smoked their first cigarette within 30 minutes of waking up. The majority of smokers (84.5%) and non-smokers (87.2%) lived with both parents. The remainder lived with either one parent or other relatives. One hundred and forty one (80.6%) of the smokers were aware that cigarette smoking is harmful to the body through the warning during advertisements (radio/bill boards). The smokers enjoyed smoking most in the following circumstances: (i) social gathering with peers/friends (60.0%); (ii) during manual work e.g., farming (15.0%); (iii) at home behind the house (11.0%); (iv) in the bedroom (10.0%) and; (v) anywhere (8.0%). A higher percentage of non-smokers (58.6%) than smokers (41.3%) expressed the desire to continue their education after secondary school. Further characteristics and attitudes of the smokers are shown in Table 2.

Table 2. Characteristics and attitudes towards smoking by 175 smokers

Characteristics and attitudes	No (%)
Reasons given by respondents for smoking	
Peer group pressure	64 (36.5)
Mix freely with peers in social situations	40 (22.9)
Imitate others	23 (13.1)
Boost self-confidence	21 (12.0)
Curiosity	15 (8.6)
Relax/Relieve anger and frustration	12 (6.9)
Opinion about smoking	
Harmful	141 (80.6)
Harmless	20 (11.4)
Uncertain	14 (8.0)
Do you smoke in the presence of your parents/guardian?	
Yes	38 (21.7)
No	137 (78.3)
Do you want to quit smoking?	
Yes	105 (60.0)
No	51 (29.1)
Uncertain	19 (10.9)
Regular source of cigarette	
Nearby shop	120 (68.6)
Hawkers	34 (19.4)
Friends	21 (12.0)
Do you send your younger siblings to purchase cigarette for you?	
Yes	53 (30.3)
No	122 (69.7)
Number of cigarette sticks smoked per day	
1 − 4	124 (70.9)
5 − 9	51 (29.1)
≥ 10	0 (0)

Characteristics and attitudes	No (%)
Do you smoke your first cigarette within 30 minutes of waking up in the morning?	
Yes	6 (3.4)
No	169 (96.6)
Academic performance of individual smoker	
Never failed	29 (16.6)
Failed once	96 (54.8)
Failed more than once	50 (28.6)

Thirty percent of smokers regularly sent their younger siblings to purchase cigarette for them (Table 2). Table 3 showed that smokers were four times more likely to have smokers among their best friends compared to non-smokers. The smoking status of fathers and elder siblings were significantly different between smokers and non-smokers (Table 3). As shown in Table 4, the highest odd ratio was found in the disapproving attitude of best friends. Higher percentage of parents and elder siblings of non-smokers disapproved of smoking by student (Table 4). Fifty (28.6%) of the smokers stated that they would like to stop smoking but their main obstacles were lack of will power, uncertainty about how to quit and fear of being regarded as timid or weakling by friends who continued to smoke.

Table 3. Prevalence of smoking among student's close contacts by smoking status of student

*Contacts	Smokers (%) n=175	Non smokers (%) n=885	Odds ratio	95% confidence interval
Father	40.1	26.5	1.84	1.23 − 2.57
Mother	0.3	0.1	0.83	0.40 − 0.90
Brother/Sister	36.5	15.3	1.76	1.11 − 2.68
Best friend	47.6	11.9	4.34	2.88 − 6.48

*In some cases, more than one students' close contact is a smoker

Table 4. Disapproving attitude of respondent's close contacts toward respondent smoking

*Contact	Disapproving (%)		Odds ratio	95% confidence interval
	Smokers n=175	Non-smokers n=885		
Parents	84.6	93.5	1.90	1.08 − 3.01
Elder siblings	68.9	84.8	1.98	1.36 − 2.89
Best friend	33.5	70.2	3.54	2.63 − 5.48

*In some cases, more than one students' close contact disapproved of smoking.

Discussion

In the present study, the overall prevalence rate of cigarette smoking (16.5%) among adolescent boys in public senior secondary school was lower than the corresponding figures reported from Saudi Arabia (20.0%) (15), Bahrain (21.0%) (17), Portugal (21.5%) (9) and Yemen (21.9%) (14), but comparable to 15.9% reported from Syria (16). In the United Kingdom, overall smoking rates have not changed appreciably over the last 20 years with 21 – 26% of adolescents admitting to smoking regularly (11). A much higher smoking rate was reported among French adolescent students (24). Differences in socioeconomic and cultural factors may account for relatively lower smoking rate in this study. The prevalence rate of smoking observed in this study was two times higher than 7.7% (19) reported among secondary school girls in South east Nigeria, indicating that cigarette smoking was commoner in boys than girls in Nigeria. The reported prevalence of cigarette smoking in general adult population in Nigeria vary from 17.6% (18) to 24.4% (1), which conformed with the report that 90% of adolescent smokers become adult smokers (8).

As in other studies (9,17,25), frequency of cigarette smoking increased with age. In this study, prevalence of smoking was higher among third-year (SS 3) than first-year (SS 1) students. The opposite was found in the study in Bahrain (17). Other available studies were silent on the issue. It is possible that the lower prevalence among first-year students in this study was because of their younger age as smoking rate increased with age in this study as well as in others (9,17,25). Another explanation might be that the intensity of peer pressure affect has not reached its peak (peer pressure effect was the commonest reason for taking up smoking in this study).

The mean age at which our respondents began smoking was 16.9 ± 1.2 years, further confirming that smoking commonly starts in adolescence. This is markedly different from 13.4 ± 2.1 years reported as mean age of commencement of smoking among Portuguese high school boys (9), suggesting that cigarette smoking starts at an earlier age in Portugal compared to Nigeria. On the other hand, the mean age (16.8 ± 1.1 years) at which smoking began in Bahrain is comparable to that found in this study. Among adult rural dwellers in South West Nigeria, the reported mean age of commencement of smoking was 18.9 ± 5.6 years (18). Compared with the present study, their higher mean age at which smoking began may be due to differences in study population (adolescents versus adults, rural dwellers versus urban dwellers). The implication of the age of commencement of smoking reported in this study is that preventive strategies should, at least, start from the junior secondary schools, or better still, from the primary school.

Majority (59.1%) of the smokers in the present study took up smoking either because they were urged on by peers and friends or simply to mix freely with peers in social situation. Available studies (9,17) among adolescent secondary school boys did not report the reason given by their respondents. In contrast, the leading reasons for smoking among Kuwaiti adults were to relieve boredom and to feel relaxed (26), two uncommon reasons among Nigerian adolescents. In the Nigerian context, therefore, cigarette smoking control programs must include equipping adolescents with skills to resist peer pressure.

In consonance with other studies (16,17,26), having parents, elder sibling(s) or best friends who smoke encouraged adolescents to take up smoking, suggesting that parents and elder siblings of smokers provided early negative role models for these adolescents. The likelihood of smoking was higher among students whose parents and elder siblings have tolerant attitude toward smoking by their sons or younger brothers. Other investigators have reported a similar finding (17,27). In this regard, for school cigarette smoking control programs to succeed, it must target not only the student, but also, their families.

Majority of the smokers in this study have repeated one or more school year due to failures, suggesting that poor academic performance promote up take of smoking. Other investigators have reported a similar finding (9,27).

This study revealed that majority (68.6%) of the smokers purchased their cigarette from nearby shops and 30.3% of the smokers sent their young siblings to purchase cigarette for them. Another Nigerian study showed that nearby shops was the commonest regular source of cigarette (18). This situation is worrisome in that it clearly shows that cigarette enjoys an excellent distribution system making procurement very easy, thereby escalating rate of smoking. The practice of sending younger siblings to purchase cigarette make them potential smokers in the near future, further fueling the cigarette smoking epidemic in developing countries.

All the smokers in this series were "light smokers" (smoked less than 10 sticks of cigarette per day). Soresi et al (28) have also noted that "light" smoking is common among high school students. It is possible that financial constraint limited the number of cigarettes consumed daily by these students. Each stick of cigarette cost between 10 and 15 Naira (approximately 0.07 – 0.1 US Dollars). However, occurrence of respiratory symptoms such as cough with phlegm and breathlessness were similar when "light" and "heavy" smokers were compared (28). They also showed that nicotine dependence can occur even in adolescent high school students who are "light smokers" (28). Only 3.4% of the smokers in this study smoked their first cigarette within 30 minutes of waking up in the morning, suggesting a low risk of nicotine dependence. Historical presence of evidence of nicotine dependence is an indication for urgent intervention program, which may include nicotine replacement therapy.

Some limitations of this study need to be considered. Failure to confirm the reported smoking status by nicotine or carbon monoxide measurements could lead to underreporting. However, other population studies have shown that data collected from adolescents by self-reporting are reliable (29,30) and once the confidentiality of the data is assured, can be accepted as valid. Caution should be exercised in extrapolating our findings to the whole population of public senior secondary school boys in Benin City since the pattern of smoking behaviour may differ from one school to another. Despite this shortcoming, the results of this study gave an insight into the magnitude of the problem of smoking among male adolescent secondary school students and highlighted the associated risk factors for smoking and the characteristics of the smokers.

References

[1]   World Health Organization. Tobacco or health: a global status report. Geneva, WHO. 1997.

[2]   Taylor AL, Bettcher DW. World Health Organisation framework convection on tobacco control; a global "good" for public health. Bull World Health Organ 2000; 78(7):920-9.

[3]   Pierce JP. International comparisons of trends in cigarette smoking prevalence. Am J Publ Health 1984;79:152-7.

[4]   Mackay J, Crofton J. Tobacco and the developing world. Brit Med Bulletin. 1996; 52 :206-21.

[5]   Warrem CW, Riley L, Asma S, Erikson MP, Green L, Blanton C, Loo C, Batchelor S, Yach D. Tobacco use by youth: a surveillance report from the Global Youth Tobacco survey project. Bull World Health Organ 2000;78(7):868-76.

[6]   Center for Disease Control. Reducing the health consequences of smoking: 25 years of progress. A report of the Surgeon General. Washington DC: US Department Health Human Services, CDC 89-841, 1989.

[7]   Shisslak CM, Crago M. Cigarette smoking. In: McAnarney ER, Kreipe RE, Orr DP, Comerci GD, eds. Textbook of adolescent medicine. Philadelphia, PA: WB Saunders, 1992:263-5.

[8]   Jenkins RR, Adger H. Substance abuse. In: Kliegman RM, Behrman RE, Jenson HB, Stanton BF, eds. Nelson Textbook of Pediatrics, 18th ed. Philadelphia, PA: Saunders Elsevier 2007:824-34.

[9]   Azevedo A, Machado AP, Barres H. Tobacco smoking among Portuguese high-school students. Bull World Health Organ 1999; 77(6): 509-15.

[10]   Working Party of the Royal College of Physicians of London. Smoking and the young. London: Lavenham Press, 1992.

[11]   Tasker RC, McChire RJ, Acerini CL. Oxford Handbook of Paediatrics. Oxford: Oxford University Press 2008:771-84.

[12]   Wheeler KC, Fletcher KE, Wellman RJ et al. Screening adolescents for nicotine dependence: the hooked on Nicotine checklist. J Adolesc Health 2004;35:225-30.

[13]   Difranza JR, Rigotti Na, McNeil AD. Initial symptoms of nicotine dependence in adolescents. Tobacco Control 2000;9:313-9.

[14]   Bawazeer AA, Hattab AS, Moraks E. First cigarette experience among secondary school students in Aden, Republic of Yemen. Eastern Mediterranean Health J. 1999; 5: 440 – 449.

[15]   Felimban FM, Jarallah JS. Smoking habits of secondary-school boys in Riyadh, Saudi Arabia. Saudi Med J 1994; 15: 438 – 447.

[16]   Maziak W, Mzayek F. Characterization of the smoking habit among high-school students in Syria. Eur J Epidemiology 2000; 16: 1169 – 1176.

[17]   Al-Hadded N, Hamadeh RR. Smoking among secondary school boys in Bahrain: prevalence and risk factors. Eastern Mediterranean Health J 2003;9: 78-86.

[18]   Ayankogbe OO, Inem VA, Bamgbala OA, Roberts OA. Attitudes and determinants of cigarette smoking among rural dwellers in Southwest Nigeria. Nig Med Pract 2003;44(4):70-4.

[19]   Ibe CC, Ele PU. Prevalence of cigarette smoking in young Nigerian females. Afr J Med Sci 2003;32:335-38.

[20]   Kokkevi A, Costas S. The epidemiology of licit and illicit substance use among high-school students in Greece. Am J Public Health 1991;81:48-52.

[21]   Ministry of Education. Directory of pre-primary, primary, junior and senior secondary institutions in Edo State. Benin City: Department Planning Research Statistics, 2006.

[22]   World Health Organisation. Guidelines for the conduct of tobacco smoking surveys for the general population, Geneva, World Health Organisation, WHO/SMO/83.4, 1983).

[23]   Ogunlesi TA, Dedeke IOF, Kuponiyi OT. Socio-economic classification of children attending Specialist Paediatric Centres in Ogun State, Nigeria. Nig Med Pract 2008;54(1):21-5.

[24]   Grizeau D, Baudier F, Allemande H. Opinions and behaviours of French adolescents confronted with tobacco in 1995. Arch Pediatrics 1997;4:1079-86.

[25]   Daniza Ivanovic M. Factors affecting smoking by elementary and high – school children in Chile. Revista de Saude Publica 1997;31:30-43. [Spanish].

[26]   Memon A, Moody PM, Sugathan TN, El-Gerges N, Al-Bustan M. Epidemiology of smoking among Kuwaiti adults: prevalence, characteristics and attitudes. Bull World Health Organ 2000;78(11):1306-14.

[27]   Lim KH, Amal NM, Hanjeet K, Mashod MY, Wan Rozita WM, Sumavni MG, Hadzrik NO. Prevalence and factors related to smoking among secondary school students in Kota Tinggi District, Johor, Malaysia. Trop Biomed 2006;23 (1):75-84.

[28]   Soresi S, Catalano F, Spatafora M, Bonsignore MR, Bellia V. "Light" smoking and dependence symptoms in high-school students. Respiratory Med 2005;99(8):996-1003.

[29]   Filsher AJ, Evans J, Muller M, Lombard C. Brief report: Test-retest reliability of self-reported adolescent risk behaviour. J Adolesc 2004;27(2):207-12.

[30]   Kentala J, Utiainen P, Pahkala K, Mattila K. Verification of adolescent self-reported smoking. Addict Behav 2004;29:405-11.

In: Public Health Concern   ISBN: 978-1-62948-424-2
Editors: J Merrick and A Tenenbaum   © 2014 Nova Science Publishers, Inc.

Chapter VIII

Smoking, smoking patterns and perceived stress in Sri Lankan undergraduates

Bilesha Perera, PhD, Mohammad R Torabi[*]*, PhD,*
Chandramali Jayawardana, PhD
and Ramani Perera, PhD
Department of Community Medicine, Faculty of Medicine,
University of Ruhuna, Galle, Sri Lanka,
Department of Applied Health Science, Indiana University,
Bloomington, Indiana, US,
Department of Natural Resources, Sabaragamuwa University of Sri Lanka,
Belihuloya, Sri Lanka and
Department of Psychiatry, Faculty of Medicine,
University of Sri Jayawardanapura, Gangodawila, Nugegoda, Sri Lanka

Abstract

Increasing prevalence of smoking among educated and high social class
people would elevate the social acceptance of the habit in the society.

* Correspondence: Mohammad R Torabi, PhD, Chancellors Professor and Dean, School of Public
Health-Bloomington, 1025 E 7th Street, Indiana University, Bloomington, IN 47405 United
States. E-mail: torabi@indiana.edu.

This chapter reports knowledge, attitudes and practices of smoking and associations between smoking and perceived health status in a sample of 1,269 undergraduates, aged between 18-28 years in Sri Lanka. A modified existing questionnaire originally developed for use in the US was adopted and used for data collection. The sample consisted of a slightly higher percentage of women (n = 731, 57.6%). Among men, 10.2% were current (monthly) smokers, and the corresponding figure for women was 1.2%, a significant difference (p<.05). Although first year female undergraduates scored higher in knowledge on smoking, compared to that of males, this gender difference of knowledge was found to be non-significant among senior undergraduates. Overall, men seem to have more favorable attitudes to smoking than women. Perceived stress was found to be positively associated with smoking in this study sample. Although the prevalence rates of smoking in this undergraduate sample is lower compared to their peer groups, the impact that undergraduates can make on society in promoting smoking is substantial. Given that tobacco industry is making aggressive efforts to promote smoking among undergraduates where the aim is to diffuse the habit to other young groups who tend to imitate their influential peers, the prevention community in Sri Lanka needs to plan and implement strategies to deglamourize smoking in educated youth such as university undergraduates.

Introduction

Smoking continues to be a vital public health issue worldwide (1,2). Prevalence remains considerably high in many countries across the world (3), and in 2000, an estimated 4.83 million premature deaths in the world were attributable to smoking (2). Not only smokers, but non-smokers as well, have become the victims of this health menace. Secondhand smoking (3,4) and ecological hazards related to tobacco cultivation (5) have caused millions of non-smokers to suffer along with smokers. Tobacco control measures taken by developed countries have forced multinational tobacco companies to look for opportunities in the developing world to sell their products. This transition made many young and economically disadvantaged people in the third world highly vulnerable to smoking. Nevertheless, a reduction in smoking in young people is observed in some developed (6,7) as well as developing (8) countries in the world. Recognition of adolescents and young people as the most vulnerable groups to initiate and continue smoking, and implementation of vigorous evidence-based preventive strategies targeted at youth are the major factors behind these public health achievements.

Sri Lanka, an island situated just beneath the southeastern tip of India, is a middle-income country. Smoking is predominantly a male habit in the Sri Lankan cultural context (9). As seen in many other poor and middle-income countries, data on smoking in Sri Lanka is scarce. A survey (N= 1565) conducted in a southern district in 2001 reported prevalence of 21% of daily smokers in males aged 18 and over (10). A slightly higher prevalence of daily smokers (24.6%) in the same sex-age group was reported by an island-wide survey (N=6698) conducted by the World Health Organization in 2003 (11). The Ceylon Tobacco Company (CTC) has the monopoly for producing cigarettes in Sri Lanka.

The CTC uses various strategies to promote smoking in young people, specially the educated young in affluent families (12,13). One such group is university undergraduates. Distribution of merchandize free of charge among undergraduates and offering them well-paid jobs in the tobacco industry (12), are some of the tactics the company uses to make a positive image of the company and their products among undergraduates. As illustrated by Social Diffusion Theory, social innovations are experimented and accepted by educated and affluent people first, and then tend to diffuse to other segments of the population (14).

The theory might well work in innovations in smoking designed by the CTC. It may be easy to popularize innovative smoking habits among other younger groups such as the unemployed, less educated youth, if a positive image of smoking can be established among educated youth. Thus, it is imperative to have a solid understanding of smoking habits in undergraduates, and personal and ecological factors related to their smoking in order to design effective control measures.

Although huge amounts of money and resources have been used by the government and non-governmental organizations to avert smoking in young people, surprisingly, very little attention has been paid to investigate the epidemiological profile of smoking in young community groups such as university undergraduates in the country. Thus, there is a need to understand patterns and related factors in undergraduate smoking.

In this chapter we investigated knowledge of, and attitudes toward smoking, and smoking habits of a sample of university undergraduates in Sri Lanka. In addition, the relationship between smoking habits and perceived levels of stress in daily life in this influential group of young people in the country was explored.

Our study

A cross-sectional survey design was employed. A self-report, anonymous questionnaire was used to collect data. Institutional review boards in the universities in Sri Lanka and in the United States approved the protocol of this study. The survey was conducted in selected classes in two universities in Sri Lanka during normal lecture hours. Three graduates were recruited and trained as research assistants to administer this study. Permission to conduct the survey was obtained from respective course lecturers. The survey was conducted in the year 2007.

The study used a modified questionnaire that was used by the second author in his previous surveys on smoking habits of college students (15). Demographic and smoking habit questions in the original questionnaire were modified or changed to suit the Sri Lankan context. For example, the question on smokeless tobacco was omitted because smokeless tobacco is simply not available for this study population. A total of 50 question items were included in the final questionnaire, among which seven were demographics, two were on personal health, 11 were on smoking habits, 11 were on tobacco use knowledge, and 18 were on tobacco use attitudes. The knowledge questions were in a multiple choice format, with one correct answer for each question, while the attitude scale consisted of five point Likert type items. The respondents' perceived stress in daily life was evaluated using one question " In general, would you say that your daily life is?, with four response categories; "very stressful," "somewhat stressful," "not too stressful," and "not stressful at all." A group of two bilingual university academics, two undergraduates, and the first author were involved in translation of the English version of the questionnaire into the native language, Singhalese, and in the back-translation. A jury of public health experts in Sri Lanka examined the questionnaire for its content, wording and appropriateness, and revisions were made based on their suggestions. The questionnaire took an average of 12 minutes to complete.

Participants were university undergraduates in two universities situated in southern and central Sri Lanka. Junior (1st year) and senior (2nd and 3rd year) undergraduate classes in the two universities were selected purposefully in order to represent undergraduates in different disciplines in the final sample. Students who volunteered to participate in the survey were surveyed using the questionnaire. Over all, about 6% of the students were from ethnic minority groups who were not fluent in Sinhalese, the local language that the majority of undergraduates speak and write, and in which the questionnaire was prepared. Therefore, those students were excluded from the survey. The overall response rate was about 95%. A total of 1294 subjects participated in the study.

Data was analyzed using Statistical Package for Social Sciences version 15.0 (16). Data was first entered in Excel data sheets and then transferred to SPSS data sheets. Internal consistency of the tobacco use knowledge and attitude items was calculated using Chronbach alpha techniques. Simple descriptive statistics were used to show variations in characteristics and smoking behavior of the sample subjects. Chi square statistics were used to identify significant categorical variables. In the analysis, those who had smoked at least once during the 30 day period preceding the survey were categorized as "current smokers" and those who had smoked at least once during the 12 months period preceding the survey were categorized as "yearly smokers." Tobacco use knowledge was evaluated using the total scores for each subject on knowledge questions, with one point for each correct answer and zero points for each incorrect answer. So the minimum and maximum score that one can obtain are zero and 11, where higher scores indicate better knowledge. Attitude items were scored from one point for the most negative (favorable to smoking) and 5 points for the most positive (against smoking) attitude. Thus, the total attitude score for each subject ranged from 18 to 90, again higher scores indicating a much healthier attitude toward smoking. Multivariate Analysis of Variance (MANOVA) was applied to detect the statistical differences in tobacco use knowledge and attitudes by gender and class standing.

Findings

Out of 1,294 subjects who participated in the survey, data on 1269 respondents were analyzed after checking for consistency of the data set. The sample consisted of a slightly higher percentage of females (57.6%). The mean age was 21.92 (SD=1.98). About 92% of the respondents were Buddhists (91.9%). In terms of marital status, the majority (98.3%) were single. Most of the students (48.5%) were from middle income families (Monthly family income 100- 500 US$) followed by poor (45.6%, monthly family income < 100 US$) and upper (5.8%, monthly family income > 500 US$). Nearly half of the respondents (46%) were junior undergraduates.

When the respondents were asked to rate their health status, nearly 75% rated themselves as very healthy or healthy, and about 24% rated their health status as average. No gender difference was found related to perceived health status of the subjects.

The reliability coefficients for the knowledge and attitude tests were 0.40 and 0.89, respectively. Knowledge test is not a norm reliability test and since it relied heavily on the criteria, its reliability is acceptable. The smoking patterns of the respondents by gender are presented in table 1.

Table 1. Smoking patterns of undergraduates by gender
(percentages are given)

		Male (n= 538)	Female (n=731)	*p value*
Any tobacco product				
	Ever	39.8	6.3	<.001
	Annually	17.1	2.1	<.001
	Monthly	10.2	1.2	<.001
	Daily	2.0	0.5	.015
Cigars				
	Ever	10.0	1.1	<.001
	Monthly	1.1	0.3	0.061
	Daily	0.4	0.0	0.099
Beedi				
	Ever	11.9	1.6	<.001
	Monthly	0.7	0.4	0.429
	Daily	0.4	0.0	0.099
Cigarettes				
	Ever	34.2	4.7	<.001
	Monthly	9.3	1.2	<.001
	Daily	1.9	0.3	.004

A clear gender difference of smoking is observed in this university undergraduate sample. The use of cigars and beedi (largely a locally produce tobacco smoking product) is not prevalent among the respondents and cigarettes seems to be the most popular smoking product. Among current smokers (those who have smoked at least once in the past month), 54.7% have tried to quit smoking at least once during their lifetime.

Out of 579 undergraduates who were from poor income families, 60 (10.4%) were yearly smokers. The corresponding figure for those who were from middle and upper income families was 6.8% (47 out of 690), a significant difference (χ^2 (1, 1269) = 5.14, p = 0.023). Further, a significant difference of the prevalence of yearly smoking was found between those who reside on-campus and those who reside off-campus. Out of 471 undergraduates who reside in on-campus hostels, 53 (11.3%) were yearly smokers, and the corresponding figure for those who reside in off-campus housing was 6.5% (52 out of 794) (χ^2 (1, 1269) = 8.59, p = 0.003).

Age of the onset of smoking by gender is presented in table 2. No significant difference was observed between male and female undergraduates in relation to the onset of smoking habits. About one- fifth of the subjects reported having first-time smoking on or before the age of 13. About half the subjects (45.9%) had their first smoking experience during the ages 14 to 17 years. Thus ages 13-17 years seemed to be the most vulnerable ages for these young people to initiate smoking.

Table 2. Age of the onset of smoking among undergraduates who have ever smoked by gender

Age (years)	Overall (n = 260)	Male (n = 214)	Female (n = 46)
≤ 13	21.2	21.1	21.7
14 -15	21.6	21.1	23.9
16 – 17	24.3	24.4	23.9
18 – 19	18.2	19.3	13.1
≥ 20	14.7	14.1	17.4

As expected, compared to non-smokers, yearly smokers scored low on average on the knowledge test (yearly smokers: M=5.86, SD=1.86; non-smokers: M=6.28, SD =1.71, t = 2.38, p<.01). Further, attitude test results indicate that on average, non smokers hold healthier attitudes towards smoking compared to yearly smokers (yearly smokers: M=65.61, SD=10.65; non-smokers: M=75.78, SD =8.50, t = 11.56, p<.01).

The comparisons of mean total knowledge scores by gender and class standing are presented in table 3. The MANOVA test showed significant

B Perera, M R Torabi, C Jayawardana et al.

interaction between gender and class standing (F=6.23, p=0.013). Even though junior females scored a higher mean value compared to junior males, this gender difference of the mean value of knowledge seemed to be diminished when it comes to senior level.

Table 3. Mean total scores of tobacco use knowledge items by gender and class stand

	Junior	Senior
Male		
n	251	287
Mean	5.87	6.34
SD	1.94	1.76
Female		
n	333	398
Mean	6.34	6.33
SD	1.72	1.52

The comparisons of mean total attitude scores of smoking by gender and class standing are presented in table 4. The MANOVA test showed no significant interaction between gender and class stand, but the differences of mean scores between gender (Male: 72.95 vs Female: 76.37, F =42.69, p<0.01) and class standing (Junior: 74.04 vs Senior: 75.67 F=8.11, p<.01) were significant.

Table 4. Mean total scores of attitudes of smoking by gender and class stand

	Junior	Senior
Male		
n	251	287
Mean	72.68	73.18
SD	8.57	10.34
Female		
n	333	398
Mean	75.06	77.46
SD	9.12	7.91

No gender difference of respondents' perceived stress of daily life was found. We further analyzed the data set to see whether there is a difference between yearly smokers and non-smokers with respect to their perceived stress of daily life. A significant difference was found (χ^2 (3, 1269) = 9.82, p = 0.02). When the respondents were asked to rate their perceived stress of daily life, 11.2% of yearly smokers compared to only 5.2% of non-smokers reported that their daily life is very stressful. The corresponding figures for somewhat stressful, not too stressful, and not stressful at all were 40.2% vs. 38.7%, 28.0% vs. 38.9%, and 20.6% vs. 17.1%, respectively.

Discussion

The findings of this study show that male undergraduates were more likely than female undergraduates to be smokers. Extremely low prevalence of current female undergraduate smokers (about 1%) confirms that the strict non-smoking norm exists in women in Sri Lanka (9,12). Although the prevalence of smoking in undergraduates is significantly low compared to such figures in many other Asian countries such as Kuwait and China (17,18) where yearly smoking figures exceed 20%, it is quite possible that the majority of those smokers who had smoked at least once during the year preceding the survey (17.1% of males and 2.1% of females) will continue the habit as they grow older and, therefore, are at risk of becoming regular smokers in the future if appropriate control measures are not taken in a timely manner. Thus, the negative impact that this group of educated young smokers could possibly pose on the Sri Lankan Society is quite substantial.

Undergraduate female smoking needs special attention in smoking control interventions. A community survey (N = 6,698) conducted by WHO in 2003, reported the prevalence of daily cigarette smoking in women aged between 18-29 year as 0% (11). In this study, a slightly higher percentage (0.3%) of female undergraduates reported daily cigarette smoking. This finding highlights the importance of monitoring and analyzing the marketing strategies used by the tobacco industry to make a smoking culture among educated women in the country as reported by Seimon and Mehl (12). Further, other factors associated with female smoking needs in-depth investigations. As expected, cigarettes is the most popular smoking product in this study population. Beedi and cigars are cheaper than cigarettes in Sri Lanka, and are smoked by people in low income brackets. Thus the two products do not pose a threat for undergraduates. However, westernization of the society and global tobacco marketing strategies may cause cigars and other "new" smoking products such as smokeless tobacco to be popular among young people in the country. The smoking control community in Sri Lanka should be vigilant on these global trends.

The results further demonstrate that smokers in this study population are more likely to have come from lower income families. This result is consistent with the findings from many other settings across the world that smoking is inversely associated with Socio-economic status (19,20). The undergraduate smokers in Sri Lanka are more likely to have resided in on-campus hostels. Studies conducted in some other countries such as Canada indicate the opposite. In a survey conducted in Canada, those undergraduates residing off campus reported higher smoking prevalence rates (21). So the campus environment seems to be conducive to smoking in Sri Lanka. To be effective, undergraduate smoking prevention programs should take into consideration these socio-economic and environmental factors.

The ages between 13-17 years were found to be the most vulnerable years for this study population to initiate smoking. Nearly 21% of those students who have ever smoked in this sample had their first smoking exposure at or before the age of 13 years. Figures for similar studies conducted in China and US were 38% and 18% respectively (15). Strict rules and regulations imposed in the US on access to tobacco products and smoking by minors may be the reason for this observed difference. In Sri Lanka, minors still have easy access to tobacco products despite recently introduced restrictions on sales of tobacco productes to minors. This atmosphere needs to be changed probably by imposing more severe restrictions to sales of cigarettes to minors.

Over half of the current smokers in the study sample have tried to quit smoking at least once during their lifetime. A cross cultural survey conducted in China and the US (15) reported lower rates of quitting smoking among undergraduate students. In the US, 45.7% of college smokers reported having tried to quit smoking and the corresponding figure for Chinese students was 23.6%. Further research is needed to explore the reasons that influence young adults in Sri Lanka from quitting smoking.

In general, as expected, the tobacco use knowledge and health attitudes towards smoking are lower among smokers compared to non-smokers. Overall, female students seem to have comparatively better tobacco use knowledge compared to male undergraduates, but as they grow older this difference seems to be reduced. This observed increase in awareness about facts related to smoking among males as they proceed to senior undergraduate levels is probably due to tobacco prevention activities in the society. Male undergraduates seem to have favorable attitudes to smoking compared to female students and this difference seemed to exist across the years of study. Peer, cultural, and environmental factors related to attitudes toward smoking need to be investigated thoroughly with regard to educated young people. Such information would be very valuable in developing strategies to destroy the social image of smoking that the tobacco industry promotes through influential population groups in the Sri Lankan Society.

Perceived stress was found to be positively associated with smoking with regard to this study population. Studies have demonstrated a higher prevalence of mental health problems in smokers compared to non-smokers in undergraduate populations (22). Poverty and other adverse ecological factors in the campus environment may contribute to make smokers' lives more stressful and finally to develop psychological health problems. Further research is warranted in this area.

There are several limitations in this study. In countries where smoking is considered as an anti-social behavior, underreporting and non-response is quite possible for questions on smoking behavior. Several efforts were made to minimize these possible errors. Whenever possible, respondents were requested to be seated in every other chair when answering the questionnaire to help them feel more comfortable in answering sensitive questions. In addition, the purpose of the survey and the importance of giving honest responses were emphasized at the beginning of the survey. Recall bias could be another limitation that usually occurs in this type of cross-sectional study. The purposive sampling method that was used limits the generalization of the results of this study. Finally, cause effect relationships cannot be determined due to the cross-sectional nature of the study.

Despite these limitations, the large sample size and inclusion of subjects from different disciplines make this study a useful research activity. The study provides important information on smoking among undergraduate students in Sri Lanka, where, according to our understanding, such information is simply not available. Findings of this study strongly support comprehensive education and health promotion programs targeted at undergraduates in Sri Lanka to maintain or decrease the currently low smoking prevalence among both men and women. Adverse effects of smoking could be more effectively addressed by tailoring and targeting preventive measures towards groups that are more at risk such as those coming from poor income families and those living in on-campus hostels. Further, findings of this study would eventually help other researchers in the region to compare epidemiological aspects of smoking in young adults in their countries, and to work together for a healthier world.

References

[1]　Britton J, Edwards R. Tobacco smoking, harm reduction, and nicotine product regulation. Lancet 2008;371:441-5.

[2]　Ezzati M, Lopez AD. Regional, disease specific patterns of smoking-attributable mortality in 2000. Tob Control 2004;13:388-95.

[3]　World Health Organization. World Health Statistics 2007. Geneva: WHO, 2007.

[4]　Warren CW, Jones NR, Eriksen MP, Asma S. Patterns of global tobacco use in young people and implications for future chronic disease burden in adults. Lancet 2006;367:749-53.

[5]　Madeley J. Tobacco: a ruinous crop. Ecologist 1986;16:124-9.

[6]　Youth Risk Behavior Surveillance System. National youth risk behavior survey: 1991-2005: trends in the prevalence of cigarette use. Atlanta, GA: US Dept Health Hum Serv, Center Dis Control Prev, 2006.

[7]　White V, Hayman J. Smoking behaviors of Australian secondary school students 2005. Carlton, Aust: Center Behav Res Cancer, Cancer Council Victoria, 2006.

[8]　Cox HS, Williams JW, De Courten MP, Chitson P, Tuomilehto J, Zimmet PZ. Decreasing prevalence of cigarette smoking in the middle income country of Mauritius: questionnaire survey. BMJ 2000;321:345-9.

[9]　Perera B. Tobacco control in Sri Lanka. Regional Health Forum 1999;3:28-34.

[10]　Perera B, Fonseka P, Ekanayake R, Lelwala E. Smoking in adults in Sri Lanka: prevalence and attitudes. Asia Pac J Pub Health 2005;17:40-45.

[11]　World Health Organization. World Health Survey, Sri Lanka 2003. Geneva: WHO, 2004.

[12]　Seimon T, Mehl GL.Strategic marketing of cigarettes to young people in Sri Lanka:"Go ahead-I want to see you smoke it now." Tob Control 1998;7:429-33.

[13]　Ranaweera S. Tobacco industry's excuses for mass murder. Ceylon Med J 1998;43:61-7.

[14]　Rogers E. Diffusion of innovations. NewYork: Free Press, 1962.

[15]　Torabi MR, Yang J, Li J. Comparison of tobacco use knowledge, attitude and practice among college students in China and the United States. Health Promot Int 2002;17:247-53.

[16]　Statistical Package for Social Sciences. SPSS for windows (version 15.0). Chicago: SPSS Inc., 2006.

[17]　Anderson JC, Palmer PH, Chou CP, Pang Z, Zhou D, Dong LI. Tobacco use among youth and adults in Mainland China: the China seven cities study. Public Health 2006;120:1156-69.

[18]　Alansari B. Prevalence of cigarette smoking among male Kuwait University undergraduate students, Psychol Rep 2005;96 (3.2):1009-10.

[19]　Siahpush M. Socio-economic status and tobacco expenditure among Australian households: results from the 1998-99 Household Expenditure Survey. J Epidemiol Commun Health 2003;57:798-801.

[20]　Xu F, Yin XM, Zhang M, Ware RS, Leslie E, Owen N. Cigarette smoking is negatively associated with family average income among urban and rural men in regional mainland China. Int J Ment Health Addict 2007;5(1):17-23.

[21]   Adlaf EM, Gliksman L, Demers A, Newton-Taylor C. Cigarette use among Canadian undergraduates, Can J Public Health 2003;94:22-4.

[22]   Gulec M, Bakir B, Ozer M, Ucar M, Kilic S, Hasde M. Association between cigarette smoking and depressive symptoms among military medical students in Turkey. Psychiatr Res 2005;134:281-6.

Section 2: Alcohol

In: Public Health Concern
Editors: J Merrick and A Tenenbaum

ISBN: 978-1-62948-424-2
© 2014 Nova Science Publishers, Inc.

Chapter IX

Predicting alcohol consumption among Chilean youth

Yoonsun Han, PhD[*1], Andrew Grogan-Kaylor, PhD[2], Jorge Delva, PhD[2] and Marcela Castillo, PhD[3]

[1]Dept. of Child Psychology and Education,
Sungkyunkwan University, South Korea
[2]School of Social Work, University of Michigan,
Ann Arbor, US
[3]Institution of Nutrition and Technology,
University of Chile, Chile

Abstract

In this chapter we estimated marginal associations of parent- and peer-related measures to examine the different patterns of lifetime ever-use and frequency of alcohol consumption among adolescents in Santiago, Chile (N=918). Probit and negative binomial models were applied to predict the probability of ever-use and the average number of drinks consumed in the past 30 days. Results supported the profound role of peer-relationships in the development of youth drinking behavior.

* Correspondence: Yoonsun Han, PhD, Department of Child Psychology and Education, Sungkyunkwan University, 25-2 Sungkyunkwan-ro, Jongno-gu, Seoul, 110-745, Korea. E-mail: yoonsunhan@skku.edu.

Particularly, peer pressure seemed more important in predicting alcohol ever-use than the frequency of drinking. Simultaneously, parents, especially fathers, played a crucial protective role. Policies aimed at preventing various drinking patterns may be more effective if they not only focus on the targeted adolescents, but also reach out to peers and parents.

Introduction

Youth-drinking behaviors are developed and manifested through social interactions with peers and parents. The role of *peers* in youth drinking patterns is critical, given the shift away from parental influence and the growing importance of friends and social relationships during adolescence. The homophily theory suggests that there is a tendency for youth to assimilate with similar-types of friends through channels of peer selection (i.e., alike individuals are drawn toward each other) and peer socialization (i.e., one influences the other) processes among individuals. Therefore, this theory maintains that peers are one of the most important elements in the development or maintenance of adolescent-related problems (1). There has been evidence in favor of both selection and socialization processes. For example, proponents of peer selection underscore a unidirectional relationship in which alcohol-drinking youth seek to meet drinking peers (2), while supporters of the socialization process assert that alcohol-drinking peers induce drinking behavior of youths (3). Furthermore, many researchers have found evidence for a bidirectional relationship, where both mechanisms of selection and socialization simultaneously link drinking patterns of youth and their peers, where alike-youths magnetize toward each other but also further develop unique drinking practices (4). All of these findings have demonstrated that peers, whether a cause or consequence, have a significant relationship with adolescent alcohol consumption.

Competing studies underscore the importance of *parenting and family* environment, such as parent-youth relationships, and parental alcohol use, in predicting youth drinking patterns (5). Social control theory proposes attachment to institutions as an element to explain youth antisocial behaviors (6). More specifically, primary mechanisms within the institution of the family that inhibit or control deviant behavior during adolescence, such as strong family bonds, propel youth in establishing greater proclivity to conform to family norms. This theory suggests that intimate family ties and support will operate as protective factors that facilitate non-drinking (or reduced drinking) despite counteractive peer (e.g., heavily drinking peers) and individual (e.g., male, older age, subject to conduct problems) risk factors. Not only is the family important as a unit, but the independent and concurrent role of mothers and fathers in preventing youth substance use have been found to be crucial, as well (7).

Recognizing the significance of both peer and parent, various statistical models have been used to identify these factors in the study of youth alcohol consumption. Duncan (1994), using latent growth curve modeling, found that family cohesion delays initiation of drinking, while peer encouragement expedites and increases the level of alcohol consumption (8). Simons-Morton (2001) suggested that peer factors may be greater than parent factors in predicting drinking and smoking behavior by computing the odds ratios in logistic regression models (9). Wood (2004) used hierarchical multiple regression analysis to examine direct associations between alcohol consumption and peer/parent influences among recent high school graduates (10). However, there has been no precedent study that estimates the magnitude of the *marginal effects* of peer- and parent-related variables alongside on youth's alcohol consuming behavior. The estimation of marginal effects may clearly illustrate the effect sizes of various measures because it can express the change in youth drinking per unit change in peer and parent variables.

As rare as studies are estimating marginal associations of peer and parent factors, there has been even less research that compares these processes for adolescent drinking in Latin America (11). The bulk of existing youth alcohol studies in Latin America has been limited to population surveys to obtain prevalence estimates with some investigation of behavioral processes. For example, a seven-country study reported that 52% of youth had the experience of consuming alcohol (12). Also, there was a greater prevalence of ever having used alcohol, and of recent-use (past year), among males and older youth in Latin American countries (13). Although lifetime prevalence of substance use among youth has been found to be lower in Latin American countries when compared to that of youth in the United States, patterns of associations between various individual characteristics and adolescent drinking patterns appear to be quite consistent across youth (12). For example, reports from the Chilean governmental organization responsible for conducting the national school and household surveys of substance use (14) and the cross-national survey of substance use coordinated by the Organization of American States Inter-American (15) have found that a greater percentage of youth with peers who use alcohol are drinkers themselves, and a smaller proportion of youth who have a positive relationship with their parents report drinking. These findings are consistent with the body of research with populations in the U.S. However, these results are based on simple bivariate analysis, and none of these, as well as other studies in Latin America (16) have examined the role of peers or parents on youth alcohol or other substance use alongside. A careful review of this body of work indicates there is a particular gap in our understanding of the potentially competing influences of peers and parents on youth alcohol use with Latin American populations.

To contribute to the study of adolescent alcohol consuming behaviors in an international context, this paper addresses two research questions using self-reported youth data from Santiago, Chile. First, what is the magnitude of peer- and parent-related measures in predicting the *probability of alcohol consumption* for youth? Second, does the importance of peers and parents change when predicting the *frequency of alcohol consumption* in the past 30 days? The hypotheses are as follows: Peers and parents will have a unique importance in determining alcohol consumption behavior when controlling for youth-specific demographic and behavioral characteristics. Also, since factors that are associated with ever consuming alcohol and those that increase the number of alcoholic drinks among youth who already drink are likely to be different, we expect the predictors of the probability of ever consuming alcohol to be different from that of the frequency of drinking.

Our study

The study, funded by the National Institute on Drug Abuse, interviewed 1,069 adolescents from municipalities of lower-middle to low-socioeconomic status in Santiago, Chile in 2007-2010. Participants were recruited from an earlier study of iron and nutritional status at the University of Chile (17). Youths completed a 2-hour interviewer-administered questionnaire in Spanish with comprehensive questions on actual alcohol/drug use and opportunities, as well as demographic, familial, peer, and neighborhood characteristics. Analysis for the present study was based 918 youths with complete data on all the variables of interest (e.g., information on both mother and father).

This study used two dependent variables, each corresponding to the two research questions. The first dependent variable was "ever used alcohol," which is a binary outcome of ever having consumed more than a few sips of alcohol ($y_i = 1$) or never having consumed more than a few sips of alcohol ($y_i = 0$) in the course of the respondent's lifetime. The second one was a count outcome that indicated the average number of drinks consumed in the past 30 days ($y_i = 0$ if no drinks were consumed). This measure was obtained by the product of the number of days alcohol was consumed in the past month and the average number of drinks per alcohol-consuming day in the past month.

The control and independent variables comprised of three domains including, youth-specific demographic and behavioral variables, parent- and peer-related variables.

Youth. *Male* was a dichotomous variable of 1 if the respondent is male and 0 if female. *Age* represented the youth's age at the time of interview. The *behavioral problems* measure was a composite score of the Youth Self-Report (YSR) problem scales ($\alpha = 0.85$) (18). The stem question of the YSR was as follows: "Below is a list of items that describe kids. For each item that describes you now or within the past 6 months, please tell me if the item is 'not true' (0), 'somewhat or sometimes true' (1), or is 'very true or often true' (2)." The study included YSR as a control variable because mental health status has been found to predict youth's substance use (19).

Parent. The family *socioeconomic status* measure consisted of thirteen questions which were combined to create a composite index score (20). This measure of socioeconomic status has been used in developing countries, and particularly, in several studies in Chile (17). These questions included information about the number of family members, head of household's market activity, father's occupation, father's education, household assets, and household utilities. A higher score indicates a lower level of socioeconomic status. *Parent's alcohol consumption* was measured by four dummy variables of "definitely no" (reference group), "probably no," "probably yes," "definitely yes" in response to a single question that asked the youth about their parents' alcohol-drinking experience during the past 12 months. The *father-youth relationship* measure (21) was the average of 17 questions ($\alpha = 0.89$) that operationalized the parent-child interaction into a four point scale ("never," "sometimes," "often," "always"). A higher value on this measure indicated that the youth's assessment of the interpersonal relationship between the father and youth was based on warmth and support rather than criticism and punishment. The *mother-youth relationship* was identified by an identical set of questions ($\alpha = 0.89$) asked in the father-youth relationship measure. An advantage of the present data set is that the father-youth and mother-youth relationship measures contained identical questions and scales, which allowed for concurrent comparison between father-youth and mother-youth relationships.

Peer. *Peer alcohol consumption* was represented by five dummy variables
– "none" (reference group), "a few," "some," "most," "all" – that measured the
degree to which friends consumed alcoholic beverages (e.g., beer, wine,
liquor). The *peer alcohol pressure* measure was a dichotomous variable that
asked the youth whether they ever received pressure to drink by friends.

Statistical model

Using the probit model, the study examined the first research question of
whether there are parent-related and peer-related marginal effects when
predicting the *probability* of youth alcohol consumption. Here, the dependent
variable was a binary outcome of whether or not the youth had ever consumed
alcohol in their lifetime. Computation of marginal effects, unlike raw
estimations of coefficients, allows the interpretation of probit results as an
expected decrease or increase in the average predicted probability of ever-
consuming alcohol with a unit change in the explanatory variable.

Second, the negative binomial count models predicted the *frequency* of
alcohol consumption. Using this model, the study investigated variables
associated with the number of alcoholic drinks consumed per month. The
negative binomial model is useful for modeling nonnegative integer outcome
values that are mostly concentrated at the lower end of the distribution (e.g.,
zero, one, and two). In the current data set, for example, 89.65 % of the youth
reported that they either did not consume any alcohol, or drank one or two
cups/drinks during the past month ($y_i = 0, 1, 2$). Marginal effects of the
negative binomial model was computed to interpret the coefficients as the
increase or decrease in the expected number of drinks consumed, in addition to
reporting raw coefficients that are interpreted as the difference in the log of
expected counts.

Findings

The average age of youth in the study sample was 14.44 years (see table 1).
There were an even number of males and females. Youth were from families
of lower-middle to low socioeconomic status and most of their parents were
consumers of alcohol.

The higher mother-youth relationship relative to the father-youth
relationship ($p<0.0001$) indicated that youths on average maintained a

warmth/encouraging relationship with fewer punitive measures and criticism with mothers, compared to fathers. The majority of youth had at least a few friends who drank alcoholic beverages and few youth reported pressure to drink from peers.

Finally, approximately 43.57% of youth had the experience of drinking alcohol in their lifetime and the average number of drinks consumed within the past 30 days was 1.79 drinks.

Table 1. Descriptive summary (*N* = 918)

Variable	Mean	Standard Deviation
Parent		
Socioeconomic Status	0.12	2.19
Parent Alcohol Consumption		
Definitely No	0.10	0.30
Probably No	0.04	0.19
Probably Yes	0.23	0.42
Definitely Yes	0.64	0.48
Mother-Youth Relationship	3.26	0.49
Father-Youth Relationship	3.20	0.53
Peer		
Peer Alcohol Consumption		
None	0.30	0.46
A Few	0.25	0.43
Some	0.25	0.43
Most	0.16	0.36
All	0.05	0.21
Peer Alcohol Pressure	0.16	0.36
Youth		
Male	0.52	0.50
Age	14.44	1.50
Behavioral Problem	44.43	19.23
Alcohol Consumption		
Lifetime Alcohol Consumption	0.44	0.50
Number of Drinks Consumed in the Past 30 Days	1.79	7.25

Multivariate results: Probit model

Table 2 reports the coefficients of three different models (parent, peer, comprehensive) using probit regression. The parent model (Model 1) showed that only the father-youth relationship (not the mother-youth relationship) had a significantly negative association with the probability of adolescent drinking. Socioeconomic status and parent's alcohol consumption were not significant predictors. The peer model (Model 2) indicated that peers' drinking behavior and peer pressure was a strong predictor of alcohol consumption. Finally, the comprehensive model (Model 3) highlighted the strength of peer variables over parent-related variables in youth's probability of ever consuming alcohol. According to the comprehensive model, the average predicted probability of alcohol consumption for SLS youth was 43.59 %, which approximated the actual SLS sample average of 43.57 % reported in the descriptive summary.

Table 2. Probability model: Predicted probability of ever alcohol consumption ($N = 918$)

Independent Variables	(1)	(2)	(3)
Parent			
Parent Alcohol Consumption[a]			
Probably No	-0.006		0.126
	(0.275)		(0.288)
Probably Yes	-0.013		-0.024
	(0.179)		(0.185)
Definitely Yes	0.190		0.187
	(0.160)		(0.166)
Mother-Youth Relationship	-0.076		-0.010
	(0.107)		(0.111)
Father-Youth Relationship	-0.234*		-0.173†
	(0.098)		(0.101)
Socioeconomic Status	0.034		0.030
	(0.021)		(0.022)
Peer			
Peer Alcohol Consumption[b]			
A Few		0.423**	0.401**
		(0.130)	(0.131)
Some		0.765***	0.739***

Independent Variables	(1)	(2)	(3)
		(0.135)	(0.137)
Most		1.145***	1.094***
		(0.165)	(0.167)
All		1.423***	1.344***
		(0.263)	(0.268)
Peer Alcohol Pressure		0.289*	0.321*
		(0.128)	(0.130)
Youth			
Male	0.271**	0.311**	0.306**
	(0.092)	(0.095)	(0.096)
Age	0.383***	0.276***	0.264***
	(0.033)	(0.036)	(0.037)
Behavioral Problem	0.008**	0.005†	0.004
	(0.003)	(0.003)	(0.003)
Constant	-5.337***	-5.142***	-4.430***
	(0.702)	(0.531)	(0.734)

Note: † $p < 0.1$; * $p < 0.05$; ** $p < 0.01$; *** $p < 0.001$ (Robust standard errors in parentheses).
[a]Reference group is "Definitely No" .
[b]Reference group is "None".

Marginal effects: Probit

Marginal effects were computed to interpret the coefficients in terms of the average probability of alcohol consumption for youth (see table 3). Consecutive analysis of average marginal effects of independent variables was based on the comprehensive model that includes both parent and peer domains (Model 3). Peers' alcohol consumption was the greatest predictor of alcohol consumption. In detail, having friends who all drink alcohol was associated with an average of 39.84 percentage points increase in the predicted probability of alcohol consumption than those with no friends who drink. On the other hand, associating with friends, "a few" of which who drink alcohol, was associated with 11.89 percentage points increase in the predicted probability of drinking relative to those with no peer-drinkers. Peer pressure to drink alcohol, on average, increased the likelihood of drinking by 9.51 percentage points. On the other hand, the father-youth relationship had a smaller marginal effect - one unit increase in the father-youth relationship measures was associated with 5.11 percentage points decrease in the predicted probability of consuming alcohol.

Table 3. Average marginal effect from Probit model (based on Model 3)

Variable	Mean	Standard Deviation
Parent		
Father-Youth Relationship	-5.12%	2.99%
Peer		
Peer Alcohol Consumption (A Few)	11.89%	3.85%
Peer Alcohol Consumption (Some)	21.92%	3.89%
Peer Alcohol Consumption (Most)	32.45%	4.60%
Peer Alcohol Consumption (All)	39.84%	7.65%
Peer Alcohol Pressure	9.51%	3.82%
Youth		
Male	9.08%	2.79%
Age	7.82%	1.00%

Multivariate results: Negative binomial

Table 4 reported the coefficients derived from three different specifications (parent, peer, comprehensive) of the negative binomial model that predicts the number of alcohol beverages consumed in the past 30 days. The parent model (Model 4) indicated that parent's use of alcohol was associated with increased frequency of drinking among youth. The father-youth and mother-youth (at a trend) relationships were also significant. In the peer model (Model 5), peer alcohol consumption was statistically significant, but the peer pressure measure was not. In the comprehensive model (Model 6), parent's alcohol consumption, father-youth relationship, and peer alcohol consumption were strong predictors of youth's average number of drinks in the past 30 days.

Table 4. Frequency model: Predicted number of alcoholic drinks consumed in past 30 Days (*N* = 918)

Independent Variables	(4)	(5)	(6)
Parent			
Parent Alcohol Consumption[a]			
Probably No	2.764***		2.090**
	(0.705)		(0.651)
Probably Yes	2.031***		1.565***
	(0.513)		(0.463)
Definitely Yes	2.025***		1.503***
	(0.474)		(0.432)
Mother-Youth Relationship	-0.596†		0.004
	(0.318)		(0.261)
Father-Youth Relationship	-0.709**		-0.733**
	(0.239)		(0.223)
Socioeconomic Status	0.051		-0.021
	(0.044)		(0.041)
Peer			
Peer Alcohol Consumption[b]			
A Few		1.213***	1.150**
		(0.354)	(0.356)
Some		2.448***	2.261***
		(0.342)	(0.350)
Most		3.882***	3.577***
		(0.379)	(0.378)
All		4.643***	4.618***
		(0.471)	(0.472)
Peer Alcohol Pressure		0.227	0.226
		(0.253)	(0.249)
Youth			
Male	0.485*	0.769***	0.664**
	(0.223)	(0.205)	(0.202)

Table 4. (Continued)

Independent Variables	(4)	(5)	(6)
Age	0.754***	0.267***	0.288***
	(0.081)	(0.072)	(0.074)
Behavioral Problem	0.011	0.006	-0.002
	(0.007)	(0.006)	(0.006)
Constant	-9.630***	-7.203***	-6.151***
	(1.659)	(1.115)	(1.442)

Note: [†] $p < 0.1$; * $p < 0.05$; ** $p < 0.01$; *** $p < 0.001$ (Robust standard errors in parentheses).
[a]Reference group is "Definitely No".
[b]Reference group is "None".

Table 5. Average marginal effect from negative binomial model (based on Model 6)

Variable	Mean	Standard Deviation
Parent		
Parent alcohol consumption (Probably No)	4.26	1.61
Parent alcohol consumption (Probably Yes)	3.19	1.15
Parent alcohol consumption (Definitely Yes)	3.07	1.08
Father-Youth Relationship	-1.50	0.61
Peer		
Peer Alcohol Consumption (A Few)	2.35	0.85
Peer Alcohol Consumption (Some)	4.61	1.09
Peer Alcohol Consumption (Most)	7.30	1.62
Peer Alcohol Consumption (All)	9.42	2.39
Child		
Male	1.36	0.53
Age	0.59	0.20

The average predicted number of drinks, using the comprehensive model, was 2.04, which is similar to the actual number of 1.79 drinks consumed among youth in the SLS sample.

Marginal effects: Negative binomial

The marginal effect of the negative binomial model was also based on the comprehensive model (Model 6). Engaging with peers who all drink was associated with 9.42 more drinks of alcohol consumption per month than having peers who do not drink at all (see table 5). On the other hand, engaging with a few peers who drink was associated with 2.346 more drinks of alcohol consumption in the past 30 days than engaging with friends who do not drink at all. Having parents who definitely have c onsumed alcohol was associated with 4.26 more drinks of alcohol consumption in the past 30 days relative to those whose parents never drunk at all. Finally, a point increase in the father-youth relationship measure was associated with 1.50 fewer drinks in the past 30 days.

Discussion

The different results between the probit and negative binomial models, suggested that predictors of the probability of ever consuming alcohol and the number of drinks may be somewhat different. By analyzing results from both probability and frequency models, the present study delineated the importance of parent and peer measures in Chilean youth's drinking behavior. The binary probit model estimated the magnitude of peer and parent factors that predict the probability of ever consuming alcohol, which encompasses a wide range of youths behaviors from one-time drinkers to heavy episodic drinkers. The negative binomial model predicted the number of alcoholic drinks consumed in the past 30 days and reported the contribution of peer and parent measures on the frequency of drinking. Since the models estimated the marginal effects of peers and parents with regards to two closely related aspects of drinking behavior, it is important to discuss the similar and dissimilar trends reported in the statistical results.

Results from both probability and frequency models indicated that the youth's relationship with the father was a significant predictor of adolescent alcohol consumption, while the relationship with the mother was not. Understanding such mechanisms through which fathers play a substantial role in predicting youth's alcohol consuming behaviors have been an under studied, yet, important topic receiving great attention (7). The current study provided meaningful results in light of the growing interest in examining the unique and independent role of mothers and fathers in the developmental literature. For example, Rohner and Veneziano (2001) reported that father's affection has been found to be a better predictor that mother's love in some studies regarding substance abuse, conduct problems, and psychological health and wellbeing (22). One explanation for the significant father-youth measure may be the relative amount of time youths spend with their mother and father. More specifically, due to the relatively less time fathers spend with their children, father's influence may be more salient than that of mothers (23). Given the growing attention placed on father's role in youth's upbringing, evidence from this study's analysis point to the need of better understanding their role, along with that of the children's mothers (24), in preventing their children from initiating the use of substances or helping them quit if already began using.

Some association between parent's alcohol consumption and youth's drinking behavior was observed in the frequency models, only. Broadly, the findings from the present study highlighted the contribution of parental drinking in predicting youth's own drinking behavior. Yu (2003) also conducted a comprehensive study that investigated the factors associated with the probability of ever consuming alcohol and the frequency of youth drinking in the past 30 days. However, results from the current analysis were not consistent with Yu's (2003) study, which concluded that parental consumption of alcohol was significant in the probability model, but not in the frequency model. The discrepancy may be due to the different sample and methods that were employed in the two studies. For example, Yu's sample was restricted to older adolescents (ages 15-18), living in New York, and used OLS to predict the frequency model (25). Therefore, there may be greater need to examine the divergent results between the probability and frequency models in the future.

The peer alcohol consumption measures were consistently significant in the probability and frequency models. This result reaffirmed the strong association between youth and peers described in the literature (4). Among all statistically significant measures, association with peers who have all had the experience of consuming alcohol had the largest marginal effect in predicting both the probability and frequency of youth drinking. This may suggest that during the unique development stage of adolescents, where interaction with peers start to become more prominent than family, the degree to which peers consume alcohol naturally becomes a strong predictor of the youth's own drinking behavior.

Interestingly, peer pressure was not significant in predicting the number of drinks (Models 5 and 6), but was significant in predicting the probability of drinking (Models 2 and 3). This may pose the possibility that peer-selection theory (i.e., birds of a feather flock together) applies for more frequent drinkers, but peer-socialization theory (i.e. flocking together, make birds of a feather) holds for one-time initiators. In other words, among more frequent drinkers it is possible that self-initiation or other factors that may present greater opportunities for alcohol-drinking rather than susceptibility to peer pressure, determines the number of drinking days in a single month. On the other hand, peer pressure may be the impetus for first initiations among vulnerable low-frequency drinkers and even among those who may never drink again.

The findings need to be interpreted within the context of two limitations. First, this study analyzed cross-sectional data, preventing us from making statements about causal inference. There may be some common characteristic of the youths, not captured in the model, that induces them to choose alcohol consuming friends and also affect their decisions whether to consume or not. Similarly, the particular parenting styles may be enforced by certain behavioral characteristics of the youth. For example, rebellious alcohol-drinking youths may compel harsh parenting measures and over-estimate the marginal effect of the father-youth relationship. Future studies would benefit from longitudinal data and statistical models that measure the developmental trajectories in alcohol use in an attempt to adjust for selection bias.

Additionally, there were potential issues with operationalization. All information collected for the analysis was based on youth self reports, the exception being the SES data. Such perceived information by youths may not actually reflect the parents' behaviors. For instance, in case of the parent's alcohol consumption measure there may be a mismatch between the parent's actual alcohol consumption and the perceived consumption by their children which may lead to a different set of conclusions. In this particular study, youth were asked whether the parents "tried" alcohol resulting in a measure with limited variability and not capturing differences in deleterious alcohol consumption. This concern also applies for measures of perceived peer behavior. Therefore, the importance of using data from multiple informants cannot be overstated. Future research would benefit from using data collected from multiple informants such as parents, peers, other adults, and teachers.

Conclusion

In conclusion, the present study is unique as it estimated and compared the different marginal associations of parent- and peer-related measures to predict the probability of youth alcohol consumption, as well as the frequency of drinking and does so with a population not generally found in alcohol studies. It is also unique in that this investigation was conducted with an international sample of Chilean youth, further contributing to our understanding of alcohol use worldwide.

By investigating these intricately related factors, we were able to depict an enhanced understanding of the determinants of adolescent alcohol behavior and their relative importance. The probit and negative binomial model results suggested the unique importance of peers and parents in youth's alcohol consumption behavior. The results do not necessarily indicate that alcohol-consuming friends cause drinking, or that the formation of friends is centered around those who are drinkers. They underline, however, that peer-relationship is profoundly associated with adolescent alcohol consumption behavior. The findings also highlighted the protective role that parents still play despite adolescents' developmentally appropriate emancipation tendencies. Finally, the role that fathers play vis-a-vis mothers in preventing their children from consuming alcohol, and in greater frequency, is a topic that merits further investigation. Policies aimed at preventing various drinking patterns may be more effective if they not only focus on the targeted adolescents, but also reach out to their peers and involve parents.

Acknowledgments

This project received support from U.S. National Institute on Drug Abuse (R01 DAD21181) and the Vivian A. and James L. Curtis School of Social Work Research and Training Center, University of Michigan. We extend a sincere appreciation to the families in Chile for their participation in this project without whom this project could not have been possible.

References

[1]   Kandel DB. Homophily, selection, and socialization in adolescent friendships. Am J Sociol 1978;84(2):427-36.
[2]   Farrell AD, Danish SJ. Peer drug associations and emotional restraint: Causes or consequences of adolescents' drug use? J Consult Clin Psychol 1993;61(2):327.

[3]   Sieving RE, Perry CL, Williams CL. Do friendships change behaviors, or do behaviors change friendships? Examining paths of influence in young adolescents' alcohol use. J Adolesc Health. 2000;26(1):27-35.

[4]   Dishion TJ, Owen LD. A longitudinal analysis of friendships and substance use: Bidirectional influence from adolescence to adulthood. Dev Psychol 2002 Jul;38(4):480-91.

[5]   Guilamo-Ramos V, Jaccard J, Dittus P, Bouris AM. Parental expertise, trustworthiness, and accessibility: Parent-adolescent communication and adolescent risk behavior. J Marriage Fam 2006;68(5):1229-46.

[6]   Hirschi T. Causes of delinquency. Berkeley, CA: University California Press; 1969.

[7]   Coley R, Votruba-Drzal E, Schindler H. Trajectories of parenting processes and adolescent substance use: Reciprocal effects. J Abnorm Child Psychol 2008;36(4):613-25.

[8]   Duncan TE, Duncan SC, Hops H. The effects of family cohesiveness and peer encouragement on the development of adolescent alcohol-use - A cohort-sequential approach to the analysis of longitudinal data. J Stud Alcohol 1994;55(5):588-99.

[9]   Simons-Morton B, Haynie DL, Crump AD, Eitel P, Saylor KE. Peer and parent influences on smoking and drinking among early adolescents. Health Educ Behav 2001;28(1):95-107.

[10]   Wood MD, Read JP, Mitchell RE, Brand NH. Do parents still matter? Parent and peer influences on alcohol involvement among recent high school graduates. Psychol Addict Behav 2004;18(1):19-30.

[11]   Latimer W, Floyd LJ, Kariis T, Novotna G, Exnerova P, O'Brien M. Peer and sibling substance use: predictors of substance use among adolescents in Mexico. Rev Panam Salud Publ 2004;15(4):225-32.

[12]   Dormitzer CM, Gonzalez GB, Penna M, Bejarano J, Obando P, Sanchez M, et al. The PACARDO research project: youthful drug involvement in Central America and the Dominican Republic. Rev Panam Salud Publ 2004;15(6):400-16.

[13]   González GB, Cedeño M, Penna M, Caris L, Delva J, Anthony JC. Estimated occurrence of tobacco, alcohol, and other drug use among 12-to 18-year-old students in Panama: results of Panama's 1996 National Youth Survey on Alcohol and Drug Use. Rev Panam Salud Publica. 1999:5-9.

[14]   Gobierno de Chile. Octavo estudio nacional de drogas en poblacion escolar de Chile 2009 [document on the Internet]. URL: http://www.conacedrogas.gob.cl/wp-content/uploads/2011/04/2009_octavo_estudio_escolar.pdf.

[15]   United Nations Office on Drugs and Crime and the Inter-American Observatory on Drugs Inter-American Drug Abuse Control Commission. Informe subregional sobre uso de Drogas en población escolarizada. Segundo Estudio Conjunto 2009/2010 [document on the Internet]. URL: http://www.cicad.oas.org/oid/NEW/Pubs/Informe_Subregional2009.pdf.

[16]   Caris L, Wagner FA, Rios-Bedoyae CF, Anthony JC. Opportunities to use drugs and stages of drug involvement outside the United States: Evidence from the Republic of Chile. Drug Alcohol Depend 2009 Jun;102(1-3):30-4.

[17]   Lozoff B, De Andraca, I., Castillo, M., Smith, J. B., Walter, T., & Pino, P. Behavioral and developmental effects of preventing iron-deficiency anemia in healthy full-term infants. Pediatrics 2003;112(4):846-9.

[18]   Achenbach TM, Rescorla LA. Child behavior checklist. Youth self-report for ages 11-18 (YSR 11-18). Manual for the ASEBA school-age forms & profiles. Vermont: University Vermont Department Psychiatry, 2001.

[19]   Garcia KSL, da Costa ML. Antisocial behavior and alcohol consumption by school adolescents. Rev Lat-Am Enferm 2008;16(2):299-305.

[20]   Graffar M. Une méthode de classification sociale d'échantillon des populations. Courrier 1956;6:455-9.

[21]   NICHD. The National Institute of Child Health and Human Development study of early child care and youth development. Research Triangle Park, NC: NICHD/RTI, 2005.

[22]   Rohner RP, Veneziano RA. The Importance of Father Love: History and Contemporary Evidence. Rev Gen Psychol 2001;5(4):382-405.

[23]   Lamb ME. The role of the father in child development. Hoboken, New Jersey: John Wiley, 2010.

[24]   Parke RD. Fathers and Families. Bornstein MH, editor. Mahwah, NJ: Erlbaum, 2002.

[25]   Yu J. The association between parental alcohol-related behaviors and children's drinking. Drug Alcohol Depend 2003;69(3):253-62.

In: Public Health Concern   ISBN: 978-1-62948-424-2
Editors: J Merrick and A Tenenbaum   © 2014 Nova Science Publishers, Inc.

Chapter X

Alcohol and drug screening of occupational drivers for preventing injuries

*Robin Christian, DNP, FNP-C, APRN**

Family Nurse Practitioner,
Alcorn State University Family Clinic,
Natchez, Mississippi, US

Abstract

The purpose of this chapter is to find out if occupational alcohol and drug testing prevents injury or work related effects. Approximately 6,000 potentially relevant publications were identified, and 19 full text articles were considered for inclusion in this review. After further review, 16 articles were excluded and two meet the inclusion criteria. Data from the two studies selected could not be statistically pooled, so a narrative description was given. From the two studies, the authors concluded that there is insufficient evidence to advise for or against the use of drug and alcohol testing of occupational drivers for preventing injuries as a sole, effective, long term solution in the context of the workplace.

* Correspondence: Robin Christian, DNP, FNP-C, APRN, Family Nurse Practitioner, ASU Family Clinic, 15 Campus Drive, Natchez, MS, 39120 United States. E-mail: rchristian@alcorn.edu.

Introduction

The research question for this review aimed to investigate the effect of alcohol and drug screening of occupational drivers (operating a motorized vehicle) in preventing injury or work-related effects such as sickness absence related to injury.

Workplace drug testing is a common procedure and/or requirement for employment, but its effect in reducing or preventing injury is unknown. Prevention and health promotion is a priority for nursing, and if an accident or injury can be prevented, the nursing community would benefit from the strategies and means to do so.

The misuse of alcohol and illicit drugs constitutes a global public health problem. Workforce drug testing is controversial and expensive. In the United States, 80% of large employers use some form of drug testing. Occupational drug testing my improve employee welfare, reduce risk to the production process, increase safety, and enhance public confidence in the company. No systematic review of the evidence has been done, prior to this study, relating to the benefits and harms of alcohol and drug testing in a workforce that primarily operates vehicles.

Study characteristics

A systematic search of the literature was done by the authors looking for randomized control trials (RCTs), cluster-randomized trials, controlled clinical trials, interrupted time series, and controlled before and after studies with an outcome measured as a reduction in injury or a proxy measure thereof. The intervention was any form of alcohol and or drug testing administered to participants with the intention of reducing work(er)-related injury. (Specific information about how the testing was completed was not given. It was stated that biological sampling and quality of testing is outlined in the federal testing regulations, to be interpreted by the management of the companies.)

Participants included; persons operating motorized vehicles in the course of commercial or public duty designed to carry passengers, persons operating motorized vehicles designed to transport goods or services, and persons operating motorized vehicles during the course of work that may not have been included under the above categories. Primary outcomes included numbers of fatal injuries, non-fatal injuries and incidents without injury.

Two interrupted time studies from the USA (Study A and Study B) met the selection criteria and were included in the review. They only included information on the schedule of data collection and on primary outcomes. Study A (N=115,019) was an evaluation of a workplace peer-focused substance abuse program and early intervention program from 1983-1996. Study A measured reportable injuries only. Study B (N was not given) was an evaluation of federally mandated random drug testing on fatal truck accidents from 1983-1997.

The primary outcome measured in Study A was the injury rate per 100,000 employee-hours. In Study B, the primary outcome measure was the rate of large fatal accidents per 100 million vehicle miles travelled. A large truck was defined as weighing over 10,000 pounds gross vehicle weight.

Data from the original papers were reanalyzed using a segmented time-series regression analysis to estimate the effect of an intervention taking into account secular time trends and any autocorrelation between individual observations. No pooling of data across studies was done because one study used measured data and the other used federal records that didn't show how many drivers the intervention affected.

Summary of key evidence

- In Study A, mandatory random and for cause testing was associated with a significant decrease in the level of injuries immediately following the intervention but did not significantly affect the existing long-term downward trend (No explanation of "for cause testing" was given and "long-term" was not defined).
- Study B showed no immediate statistically significant effect for mandatory testing.
- For long-term change (again, not defined), Study A's interventions found a significant decline of the yearly injury rate and Study B's interventions found a significant improvement in the downward trend (not defined).
- Study A showed a drop of 0.19 injuries per 100 persons and Study B found that mandatory drug testing can increase the downward trend of the rate of fatal accidents by -0.83 fatal accidents per 100 million vehicle miles traveled per year.

- Study A also found that drug and alcohol testing interventions may be cost effective in the United States.

Best practice recommendations

Limited evidence was provided in the two studies that mandatory random drug testing can decrease injuries in the long-term although the results in the short term were contradictory.

Since there is insufficient evidence to advice for or against the use of alcohol and drug testing for occupational drivers for preventing injuries, the authors recommend the need for cluster randomized trials to better address the effects of interventions for injury prevention in the occupational setting. A cluster-randomized trial would be the ideal study design to evaluate the effects of interventions for injury prevention in the occupational setting.

References

[1]   Cashman CM, Ruotsalainen, JH, Greiner BA, Beirne PV, Verbeek JH. Alcohol and drug screening of occupational drivers for preventing injury. Cochrane Database Syst Rev 2009;2:CD006566.

Section 3: Other substance use

In: Public Health Concern   ISBN: 978-1-62948-424-2
Editors: J Merrick and A Tenenbaum   © 2014 Nova Science Publishers, Inc.

Chapter XI

Transitional drug use: Switching from alcohol disability to marijuana creativity

Hari D Maharajh[†], Jameela K Ali and Mala Maharaj[*]

Psychiatry Unit, University of the West Indies, Mt Hope Hospital,
Trinidad, Mt Hope Medical School, University of the West Indies,
Trinidad and St Anns Hospital, Trinidad, West Indies

Abstract

The raging controversy of the benefits of marijuana use over that of alcohol is far from over and will be on the front burners for some time in the future. The debate has been a two-way tug-of-war with each proponent submitting strong arguments, albeit, at times emotively charged. Those involved in the dual use of marijuana and alcohol have experimented among themselves and some have made choices of a single drug maintenance use that enhances their lifestyle, personality and

---

[†] Deceased.
[*] Correspondence: Professor Joav Merrick, MD, MMedSci, DMSc, Medical Director, Health Services, Division for Intellectual and Developmental Disabilities, Ministry of Social Affairs and Social Services, POBox 1260, IL-91012 Jerusalem, Israel. E-mail: jmerrick@zahav.net.il.

productivity. A few have attained a steady state while many continue to live disorganized lifes of chaos and confusion. Transition drug use (TDU), that is, the replacement of alcohol abuse with its addictive properties with the non-addictive cannabis has been attained by some polydrug abusers. We wish to present a case report presentation of a 45 year old female alcoholic who made a successful transition from alcohol dependency to a marijuana maintenance programme (MMP). Over a period of six years, she attained stable-state social and occupational adjustments with her own financial support and productivity. It appears that a marijuana maintenance programme can be utilized in carefully selected individuals as a replacement for alcohol abuse disorders. Transitional drug use to marijuana offers the opportunity of episodic use, less effect on behavior, more personal interest in creativity, absence of bulk and less need for group socialization.

Introduction

The marijuana maintenance programme has been adopted by some alcohol dependent individuals who substitute the use of marijuana (cannabis) as a replacement for their alcohol dependency. There are cogent and convincing discussions for and against the use of marijuana maintenance. It is generally accepted that the use of marijuana may indeed be less hazardous than heavy drinking among those who have been drinking excessively for a long period of time. On the other side, the excessive use of uncontrolled marijuana use is associated with negative effects and consequences: this encompasses its illegality in certain jurisdictions, its effects on mental health, learning and social withdrawal and even psychosis (1).

Cannabis and its most active ingredient delta-9 tetrahydrocanabinol (THC) are derived from the plants Cannabis indica and cannabis sativa which were transported to Trinidad during the indentureship period between 1845 to 1919 by East Indian emigrants who were imported to work on the sugarcane plantations. Locally referred to as marijuana, ganja, hashish, pot and weed, the substance was originally intended for medicinal purposes but due to its hallucinatory and euphoric effects has found a place in creativity, religion and as a street drug of abuse. The lethality of this drug is increased when the smoke is inhaled directly, facilitating easy entry into the blood stream and quick absorption into the brain.

Substance abuse is a worldwide problem with increasing use of tobacco, alcohol and cannabis among children and adolescents in the school systems.

Internationally, the patterns of use are changing among youths, with an increasing use of cannabis (2) and alcohol (3,4). Lifestyle changes associated with the use of these drugs have resulted in more social problems (4) and criminal violence (5). Marijuana today, has found a niche among young children from the ages of twelve years and has overtaken alcohol as the primary drug of use in many inner city schools.

Both alcohol abuse and cannabis consumption are common problems in Trinidad with high prevalence rates in both genders. Over a period of 15 years, a threefold increase from 8% in the rates of cannabis use has been observed in Trinidad and substance use has been directly related to crime and suicidal behavior (6,7). Worldwide, the age of initial use has decreased to a mean of 12 years with the rates of cannabis use among adolescents who have used the substance at least once in their lifetime ranging from 32.5% to 43.0%. The lifetime incidence ranges from 3.5% to 8% cannabis use to 2.2% to 7% for continuous cannabis use sometimes termed dependency. The incidence rates of alcohol, tobacco and cannabis use was even higher in youths with disabilities who reported significantly more exposure to risk factors and fewer protective factors. Studies in developing countries have also reported high rates of cannabis and alcohol use.

The Caribbean society is faced with the problem of early induction into substance abuse. Ganga tea, referred to as 'the elixir of life' is commonly used in Jamaica, while in Trinidad, alcohol based beverages are the gateway drugs. There is an ethnic preference to these drugs, those of African descent having a preference for marijuana while those of Indian origin choosing alcohol. This pattern continues into adult life. Observations on the functional ability of the use of these two substances in the local settings suggest that marijuana may enhance the creative abilities of certain individuals. On the other hand, the pattern of alcohol abuse can dampen creativity, due to the patient's loss of control or inability to abstain. Further research is needed n this area. Drug tourism is common in many Caribbean countries including Costa Rica and legal authorities often turn a blind eye to marijuana use. In these countries, it is a statutory illegal drug with 'legal sanctions' of social use.

Case report

This is the presentation of her personal lifestyle and thirty years of drug use by Karen S.

Marijuana is one of the most studied and controversial substances currently with millions being pumped into research of the effects of the drug use. In some countries it has been legalized, however, in the majority of the world it continues to be illegal and there is a stigma associated to those who chose to use it.

Karen S is a Trinidadian female in her mid 40's who currently resides in Costa Rica. She is a classical musician and has kindly shared her personal experience with marijuana use. She grew up in a family, where she was surrounded by alcoholics. These included her grandfather, aunts, uncles and most influentially, her father. She saw day by day, consumption of large quantities of alcohol and the consequences, which the most significant being the death of her grandfather at age 42 years of alcoholic liver cirrhosis and her uncle at age 50 years with the same fate. She described her childhood as happy and relatively normal. She described herself as being 'bright but disobedient'. She elaborated that at a young age she had always been very creative and was not a typical child, always being more expressive.

Karen S first experience with alcohol occurred far before she developed memories. The event was described to her by her mother many years later. Her mother said she was quite an irritable baby and she had difficulty in getting her to sleep. Relatives advised her mother that some brandy in her formula would help. Her mother apparently did not use the brandy in discretion and the quantity was so much that she actually threw up.

She attended the Montrose APS Elementary Vedic School. Though a difficult child, she performed well academically and passed for the prestigious Bishop's Anstley High School, when she wrote the qualifying examination. She recalled doing well in school, however, by Form 5 (age 16 years) had begun to consume alcohol regularly and some friends and her would skip classes to 'go drink'. Due to this, her academic studies faltered and she was no longer interested in studying. Nearing her seventeenth year matriculation from high school, she said she felt everyone in her family expected her to do well since they did not know about the extent of her alcohol consumption. Knowing she was unable to fulfill their expectations she refused to write the exams.

At 17 years, Karen S family sent her to New York to study hairdressing. She said this allowed her an unrestricted atmosphere that only allowed her to consume alcohol more frequently. A few years later, Karen S got married and according to her, she had a pretty good marriage, a good home and soon after twin boys. Though this should have made her happy and give her a sense of accomplishment and contentment she said she often experienced bouts of severe depression, where she secluded herself, felt insecure, anxious and lonely. She claims in these times she drank heavily, because when she drank she felt more outgoing, confident and popular. Unfortunately, her drinking soon led to divorce. She recalls after the divorce, she continued to drink heavily and started to have relationships with men, who drank similarly. All these relationships were short term and tinged with promiscuity.

Karen S moved to Costa Rica nine years ago . She met a man with similar musical interests and they started a band together and also a romantic relationship. However, he like boyfriends previously was a chronic alcoholic. She recalled that it was in this relationship she had somewhat of an epiphany, where, she finally saw the effects of alcohol addiction. She says she always knew it was not a good habit and she knew it was damaging her body and she would have to stop eventually, but only in this relationship she saw firsthand the habits of a chronic alcoholic. She said her boyfriend at the time was a good man and a talented artist, however, he was unable to function without alcohol everyday and it was at this point she gave him an ultimatum: it was the alcohol or the relationship.

He chose the relationship and she acted like his nurse and support system during his withdrawal period. She experienced first hand his withdrawal symptoms such as nausea, irritability and hallucinations. Karen S had experimented with marijuana as a teenager, but had experienced palpitations and did not like the feeling she got. During this period of alcohol withdrawal with her boyfriend she abstained from alcohol as well to be supportive and additionally she did not want to reach the point of alcohol necessity and the symptoms of alcohol withdrawal. It was here that she experimented with marijuana again to calm her anxiety.

Karen S stated that she had visited four psychiatrists previously for anxiety and depression and was prescribed different medications. Although she admitted the medicine worked, they produced many undesirable side effects such as lethargy and weight gain, so she stopped using them. On her own accord, she embarked on a marijuana maintenance program where she no longer used alcohol, but instead marijuana for her anxiety. For six years Karen S continued the relationship, however, her boyfriend started to drink occasionally and she said she knew the habit would evolve to addiction again so she ended the relationship.

Karen S joined an Alcoholics Anonymous Group for support and though she does not consume alcohol anymore she still used marijuana. She does not see a conflict of interest since the philosophy of AA is a desire to stop drinking. Currently, she still uses marijuana in controlled amounts and stated clearly that she does not abuse it. She admits that she will turn back to alcohol if she stopped using marijuana but with it she is able to live comfortably. She said she is able to make better decisions, she commands more respect, people come to her for advice and she is able to be supportive to others with alcohol addictions to name a few. These characteristics were completely absent while she drank.

Karen S is a classical musician and she said when she smokes she becomes introverted and is able to think clearer. She is able to be creative and focused and accomplish tasks she set out to do. She said it brings out her artistic ability and helps her relax. She states she is not religious, but rather spiritual and when she smokes she feel 'at one with her Maker'.

She stated clearly that she uses marijuana in moderation and mainly to practice. She does not use it during performances or for day to day interactions because she said that in these instances she wants to be out-going and interactive and marijuana makes her more introverted. She insisted that she was not here to force her vice on anyone but only wanted to share her personal experience. She also stated that she would not recommend it for everyone but it worked for her. The only withdrawal symptom she experiences is mild cold sweats for a couple of days which in her opinion is much better than alcohol withdrawal symptoms. All in all Karen S said of marijuana the much clichéd statement "I like it and I am glad it grows on the Earth! It has been here for a millennium!

Discussion

Slavery and indentureship, not unlike the holocaust in Europe are important driving forces of human behavior in the Caribbean. In the local setting, this is linked to alcohol and drug abuse.

The advent of indentured workers from India to the Caribbean following the emancipation of slavery in 1838 resulted in the introduction of the plant Cannabis indica. Cannabis referred to as marijuana, weed, ganja, herb, hashish, bhang or pot has many ingredients, the most well known being delta-9-tetra hydrocannabinol.

It is prepared from flowers, leaves, stems and seeds of the plant, can be smoked, liquefied into teas and tonics or grounded and baked in cookies and cakes. Its earliest use was intended for adjustment of East Indians into a new and hostile environment, separated from their loved ones thousands of miles away, oppressed by the colonial masters and socially rejected by the freed slaves who perceived them to be strange and illiterate. It was used as a balm for solitude and quiet reflection and medicinally for stress related disorders, asthma, pain relief, arthritis, malnutrition and other ailments.

Historically, cannabis use in the Caribbean was endemic with good social adjustment to its consumption. Until 1950, it was a legal drug that could be bought in shops and parlors. This was changed for economic and social reasons. The early pattern of collective use for recreation, enlightenment and socialization gave way to individual high quantity use for the sole purpose of "building a head". This changing pattern of use also created a lucrative market for the drug trade. It was soon transformed into a drug of abuse due to loss of cultural sanction in its usage.

Alcohol abuse, on the other hand is also endemic in its use in Trinidad and Tobago. It is the drug of choice and addiction of the migrant East Indian population that was transported 162 years ago to work on the sugar plantations. It is the most popular drug of use at home and even given to babies as Karen S pointed out for sedation and sleep. Alcohol is the most commonly used drug at schools and the one that parents and authorities object to least.

Karen S was raised in an alcoholic family and as an adolescent experimented with alcohol. Despite her apparent above average ability, she performed poorly at school, not writing her examinations, having an early marriage with children, marital divorce and migration to Costa Rica where she established a number of unstable relationships with mostly substance abusers recapitulating her family behavior. Through role reversal with an alcoholic partner, she was able to change her lifestyle by switching to the controlled use of the less addictive marijuana. This worked well for her, enhancing her creativity and decreasing her promiscuity of one night stands so commonly found among alcoholics.

Her method of use of marijuana is smoke inhalation through cigarettes which enabled her to monitor her blood levels. She felt the eating of marijuana baked cookies did not allow for predictable control due to the absorption process. At the end of her presentation, a number of questions were posed to Karen S:

- **Does you AA group know about your addiction?**
  Yes, they do. Karen S stated that she is 45 and will not lie to fit in or to please others. She informed that the only requirement for AA is a desire to stop drinking and the 12 step program does not include anything about marijuana use. Some group members agree with her decision and others disagree. She is obviously closer to those who agree with her decision,

however, she does not discuss or impose her views on marijuana during AA meetings.

- **Are you addicted?**
  Karen S does not believe she is addicted as she views an addiction as something that will ultimately lead to illness or death and she said marijuana causes neither. She said she wishes she did not have to use anything but is glad there is something that keeps her away from alcohol. She also stated that she has been on vacation to Trinidad for the past two weeks and has not used any marijuana during this time and she feels fine and has experienced no major withdrawal symptoms. She emphasized the view that she used marijuana mainly for the practice of her music and creation of miniature life-like dolls.

- **Does your dosage increase?**
  Ms. S. said that there are different qualities of marijuana so different amounts might have to be used to get the same effect. She said that any drug can be abused and marijuana is sometimes mixed with cocaine to produce addiction and some people use marijuana in combination with other drugs and falsely blame the ill-effects on the marijuana. She said if marijuana is being used no other drug should be used. Personally, she said her dosage has remained constant.

- **How does it affect you negatively?**
  Cost of marijuana and people's attitude to the use of marijuana. She has no problem with the law makers in Costa Rica, who turned a blind eye to her personal use.

- **Can you use it as a tea?**
  You can make cookies, cakes, teas or be used in a vapourizer so you will not have to smoke and create any lung damage.

- **How does marijuana affect your sexual behavior compared to your former use of alcohol?**
  In her experience when she consumed alcohol she was more promiscuous, with more than one boyfriend at any given time and she was not very honest or fair to her partners. She was less organized, sleeping out, empty and void. With marijuana, she has been celibate for

about 18 months and has no desire for multiple partners. Her sexual interests and promiscuity has certainly diminished.

- **Does marijuana make you depressed sometimes?**
  No. Karen S said she experienced depression while she was an alcoholic but does not experience those symptoms anymore even if she does not smoke. Marijuana does not make her depressed.

- **If there is nothing wrong with it, why not recommend it to everyone?**
  Karen S said she does not believe that everyone needs it.

- **Does it prevent sleepiness?**
  Karen S is not sure but did say it affects people differently.

- **What is the difference between the cookie marijuana and smoking it?**
  She recommends that it can be eaten to prevent the effects of smoke in the lungs but said that she prefers to smoke it because she can control how much she uses better. She said that in cookie form the effects are felt an hour later and if you eat too much you can't do anything to reverse the effects at that point.

- **Do you think it should be legalized?**
  Ms. S. believes that it should. She said that she believes that every drug can be abused and has disadvantages but she believes that the advantages of marijuana far outweigh the disadvantages.

Conclusion

The Caribbean today is a major area of cannabis consumption and trade. There are problems of control despite assistance for drug interdiction by the United States Government. Cushioned between the vivid bio-psychosocial effects of alcohol abuse and criminal culture of cocaine, cannabis is often viewed as a harmless drug. This perception may have emerged because of its presence for more than century and a half with clear cut social sanctions of its use. These sanctions of consumption are no longer tenable since the pervasive youth culture has devised rules of its own. Consequently, cannabis use among adolescents is on the rise with increasing adverse effects. Cannabis remains the drug of choice for the modern day Caribbean youth merging without suspicion into educational, creative and sporting activities without the burden of bulk.

Karen S transition from alcohol dependency to a stable state MMP is not unique, especially with the use of the twelve steps and traditions of AA which are now applicable to a wide variety of disorders. Attempts to structure a marijuana maintenance programme (MMP) for alcohol withdrawal must be carefully planned. Variables of age, intelligence, ethnicity, gender, premorbid personality, level of alcohol disability, support systems, family history and the mental status must be taken into consideration. This is no easy task.

TDU provides an opportunity for a group of drug addicts to effect lifestyle changes through changes in their drug taking habits. The psychodynamic of this change has not been fully worked out, whether it is a maturation process, personality factors or social situations. It is noteworthy that a MMP finds support in the London Drug Dependence clinics where scripts are given to addicts for maintenance therapy. While Alcoholism has been dubbed a bio-psychosocial disease that ravages families, a similar virulence has not been observed with marijuana use. On the other side, marijuana use is illegal in the Caribbean and US drug interdiction intervention results in fines and imprisonment. There is no legislation for personal use and this carries the burden of imprisonment. As with Karen S, policemen often turn a blind eye more often to female users. It should be noted that the authors do not support the de-legalization or decriminalization of cannabis use and do not view it as an innocuous drug. We do not share the optimism of Karen S. Further research is needed.

References

[1]   Buddy T. The dangers of the marijuana maintenance program substituting smoking pot for drinking alcohol, about.com guide. Updated November 11, 2010.

[2]   Rossow I, Groholt, Wichstrom L. Intoxicants and suicidal behaviour among adolescents: changes in levels and associations from 1992 to 2002. Addiction 2005;100(1):79-88.

[3]   Berggren F, Nystedt P. Changes in alcohol consumption. An analysis of self-reported use of alcohol in a Swedish national sample 1988/89 and 1996/97. Scand J Public Health 2006;34(3):304-11.

[4]   Poelen FA, Scholte RH, Engels RC, Boomsma DI, Willemsen G. Prevalence and trends of alcohol use and misuse among adolescents and young adults in the Netherlands from 1993 to 2000. Drug Alcohol Depend 2005;79(3):413-21.

[5]   Javier Alvarez F, Fierro I, Carmen del Rio M. Alcohol-related social consequences in Castille and Leon, Spain. Alcohol Clin Exp Res 2006;30(4):656-64.

[6]   Haggard-Grann U, Hallqvist J, Langstrom N, Moller J. The role of alcohol and drugs in triggering criminal violence: a case-crossover study. Addiction 2006;101:100-8.

[7]   Singh H, Maharajh HD, Shipp M. Pattern of substance abuse among secondary school students in Trinidad and Tobago. Public Health 1991;105:435-41.

[8]   Maharajh, HD, Koning, M. Suicidal behavior and cannabis-related disorders among adolescents. In: Merrick J, Zalsman G, eds. Suicidal behavior in adolescence. An international perspective. Tel Aviv: Freund, 2005:119-29.

[9]   Ali A, Maharajh HD. Social predictors of suicidal behavior in adolescents in Trinidad and Tobago. Soc Psychiatry Psychiatr Epidemiol 2005;40:186-91.

[10] Konings M, Maharajh HD. Cannabis use and mood disorders: patterns of clinical presentations among adolescents in a developing country. Int J Adolesc Med Health 2006;18(2):221-33.

In: Public Health Concern ISBN: 978-1-62948-424-2
Editors: J Merrick and A Tenenbaum © 2014 Nova Science Publishers, Inc.

Chapter XII

Substance use and the workplace: Adolescent and young adult employees

Jessica Samuolis, PhD[*1], *Kenneth W Griffin, PhD*[2],
Christopher Williams, PhD[3], *Brian Cesario, MA*[4]
and Gilbert J Botvin, PhD[3,4]
[1]Department of Psychology, College of Arts and Sciences,
Sacred Heart University, Fairfield, Connecticut
[2]Division of Prevention and Health Behavior, Department of Public Health,
Weill Cornell Medical College, New York
[3]National Health Promotion Associates, White Plains, New York,
[4]Inwood House, New York, New York

Abstract

In this chapter we investigated the relationship between number of hours
worked, or work intensity, and substance use in a sample of adolescent
employees of a supermarket chain. Employees working half-time or more
per week (high-intensity hours) were over three times as likely to smoke

---

[*] Correspondence: Jessica Samuolis, PhD, Sacred Heart University, Department of Psychology,
College of Arts and Sciences, HC 219, 5151 Park Avenue, Fairfield, CT 06825, United
States. E-mail: samuolisj@sacredheart.edu.

compared to those working an average of 10 hours or less per week (low-intensity hours). Males working a high intensity number of hours were more than twice as likely to drink compared to males working at low intensity. Utilizing participants drawn from a uniform employment setting, the research findings add to the growing body of evidence linking work intensity with adolescent substance use.

Introduction

During the adolescent years, a large number of youth are employed in a variety of work settings. Potential benefits of part-time employment include gaining valuable work experience and learning how to handle new responsibilities. Work experiences may also facilitate a sense of independence, identity development, and promote other developmental goals of adolescence. However, for some youth employment may provide increased opportunities to engage in risk behaviors. In fact, evidence is accumulating that there is a positive association between employment and substance use among young people in the workforce. Work intensity, the number of hours worked, has been found in several studies to be positively associated with tobacco use (1-4) and alcohol use (1,2,4,5).

A number of additional studies have identified subgroup differences in the relationship between work intensity and substance use. Some research has examined whether this relationship depends on the specific substance examined. Although the findings are mixed, there is some evidence that the link between work intensity and alcohol use is stronger than the corresponding link with tobacco use (6). Other researchers have identified differences in the relationship between work intensity and substance use according to race/ethnicity and/or gender. Johnson (7) found that more hours worked was positively associated with substance use among white adolescents, but was not consistently related among minority adolescents. A small number of studies have examined gender differences, finding hours worked to be associated with cigarette smoking for white males, but not for white females (2), or a weaker association between work intensity and alcohol use for females, compared to males (1).

In order to fully understand the relationship between work intensity and substance use, it is important to recognize that workplaces and job types are likely to differ in a number of relevant ways. Indeed, as stated by Zimmer-Gembeck and Mortimer (8) "The experience of adolescent work, including tasks and responsibilities, rewards, workplace relationships, and the meanings of these experiences, take fundamentally different forms in distinct historical, sociocultural, institutional, and social class contexts" (p. 538). Some work contexts provide the opportunity to develop work-related technical skills, interpersonal skills in areas such as management and conflict resolution, and/or provide extensive adult mentorship and supervision, and these positions may be associated with less antisocial behavior (9). Some youth that work in family-owned businesses have been found to use substances less than others (10). On the other hand, many employers hire large numbers of young people for unskilled or low-skilled labor for relatively brief periods of time and therefore may be less likely to provide training, mentorship, or adult supervision. Adolescents whose work experiences do not provide opportunities for skill development and/or access to adult role models or mentors, such as in the low-wage retail and service positions commonly available to adolescents, may be at greater risk for substance use. For example, youth employed in low-wage, service sector jobs have been found to have a heightened risk for delinquency (11).

In light of this, it is important to more closely examine not only the extent of employment and its relation to adolescent problem behavior, but also investigate this relationship in particular workplace contexts. A limitation of most studies on this topic is that they draw from large national databases (i.e., National Longitudinal Study of Adolescent Health) and/or school-based samples, in which youth are employed in a variety of industries and job types. A goal of the present study is to examine the relationship between work intensity and substance use in a particular work setting/industry among employees of a common job type – that is, entry-level, low-skilled, part-time positions in a large supermarket chain. This paper examines the relationship between work intensity and cigarette and alcohol use in this sample, as well as potential gender differences in this association.

Our sample

The sample included 648 adolescent and young adult employees at a large supermarket chain in the northeast. All subjects were between the ages of 15 and 20 years old, with a mean age of 18.38 (SD=1.13). More than half 344 (53.1%) of the subjects were male and 304 (46.9%) were female. The majority of the sample (94.6%) was not of Hispanic/Latino descent, and included the following racial characteristics: 91.8% White, 6.3% Black/African American,

3.5% American Indian, 1.1% Native Hawaiian or other Pacific Islander, 1.7% Asian, and .3% Alaska Native.

The majority of the subjects 508 (78.4%) reported being currently enrolled in school either full-time or part-time. Education levels of participants were as follows: 351 (54.1%) less than high school degree; 142 (22%) highest level of education was a high school degree; 97 (15.0%) coursework after high school but no degree; 12 (1.8%) obtained an Associates or Bachelor's degree; 2 (.3%) reported graduate coursework but no degree; and 44 (6.8%) did not provide data on education level. The majority 587 (90.6%) of the subjects were never married and 597 (92.1%) lived at home with a parent(s) or foster parent(s).

Participants worked in urban (33.3%), suburban (37.8%), and rural (28.9%) worksites (N=20). Participants worked part-time and held positions such as grocery bagger, grocery cashier, grocery stocker, produce clerk, and food preparation worker. Although data on income was not collected, these positions are traditionally considered entry-level and low-wage. Additionally, these positions require few skills and offer limited access to adult mentors. Employees who held managerial roles (i.e., department manager, assistant manager) were not included in the present sample. Approximately half of the sample (48.3%) worked at their current position for less than one year.

The sample examined in this study was part of a larger study that included young people up to the age of 24 years. In this paper, we restricted the sample to those subjects under 21 years of age in order to focus on predictors of smoking and underage drinking. Those participants that provided information on gender, date of birth, substance use, and the numbers of hours worked in the past month were included in the present analysis.

Study participants were recruited as part of a randomized control trial evaluating a workplace-based substance use prevention program. The data for the present study were drawn from the baseline data collection, prior to any intervention. Project field staff recruited subjects using posters at the worksites and flyers attached to employees' paychecks. Field staff distributed and collected consent forms and surveys on site at each worksite. All subjects completed consent forms and parental consent was obtained for minors. The consent form included a description of the larger study, which involved attendance in a skills-based wellness workshop and the completion of several follow-up assessments. Additionally, the consent form indicated that participation in the study would not impact employment status and that individual survey responses would not be shared with their employer or parents. All procedures were approved by an Institutional Review Board.

The twenty-minute, paper-and-pencil survey included standard demographic items and scales assessing substance use, work variables, and intervention specific variables. Demographic items included age, gender, race, ethnicity, marital status, education level, and household composition. In order to assess lifetime use of cigarettes and alcohol, participants were asked to indicate yes/no to "Have you ever smoked all or part of a cigarette?" and "Have you ever, even once, had a drink of any type of alcoholic beverage?" To assess past 30-day substance use, participants were asked "How long has it been since you last smoked part or all of a cigarette?" and "How long has it been since you last drank an alcoholic beverage?" with response options of within the past 30 days, more than 30 days ago, but within the past 12 months, and more than 12 months ago. Similar items were used to assess smokeless tobacco use, marijuana use, and the use of other illicit drugs, although this data is not included in this study.

Demographic items and substance use items were drawn from well-established tools such as the National Outcome Measures (NOMS), the Government Performance and Results Act (GPRA), and the National Survey on Drug Use and Health (NSDUH). Subjects were also asked to indicate the number of hours worked in the past four weeks along with other items from the World Health Organization Health and Work Performance Questionnaire (HPQ; Kessler et al., 2003). There is little consensus in the literature regarding the classification of work intensity. Some researchers have used a cutoff of 20 or more hours worked per week (5, 13), others have categorized hours worked in five-hour increments (2, 14), and others have categorized hours worked in ten-hour increments (3, 4).

In light of some research suggesting that working ten hours or less per week may be protective in terms of risk of substance use (3), work intensity is classified here as "low-intensity" for those working 10 or less hours per week (1-40 hours per month), "moderate-intensity" for those working more than 10 but no more than 20 hours per week (41-80 hours per month), and "high-intensity" for those working 21 or more hours per week (81-140 hours per month) as a part time employee.

Findings

Chi-square analyses were run to examine the prevalence of lifetime and past 30-day cigarette and alcohol use among male and female participants. As indicated in table 1, females had higher frequencies of lifetime cigarette use chi-square (2, N = 146) = 5.84, p < .05 and alcohol use chi-square (2, N = 229) = 9.97, p < .01 than males. Although not significant, this gender difference existed for past 30-day cigarette use, but not for past 30-day alcohol use.

Table 1. Prevalence of lifetime and past 30 Day cigarette and alcohol use

Substance	Gender		Chi-square
	Males [N=344]	Females [N=304)]	
Lifetime			
Cigarettes	133 (38.7)	146 (48.0)	5.84*
Alcohol	219 (63.7)	229 (75.3)	9.97**
Past 30 Day			
Cigarettes	59 (17.2)	70 (23.0)	3.49+
Alcohol	116 (33.7)	118 (38.8)	1.82

An additional chi-square analysis was run to determine if there were any significant differences in the frequency of number of hours worked in the past four weeks between males and females.

Self-reported number of hours worked in the past four weeks was collapsed into a categorical variable with three levels (1-40 hours, 41-80 hours, and 81-140 hours). There were no statistically significant differences chi-square (2, N = 146) = .89 (see table 2).

Table 2. Hours worked past month

Hours	Gender	
	Males [N=344]	Females [N=304]
1-40	99 (28.8)	97 (31.9)
41-80	131 (38.1)	107 (35.2)
81-140	114 (33.1)	100 (32.9)

Binary logistic regression analyses were used to predict male and female participants' current cigarette and alcohol use based on number of hours worked in the past four weeks. The dependent variables used for these analyses were dichotomous self-report past 30-day cigarette and alcohol use scores. These variables were coded as either 1 for reported use or 0 for non-use. As seen in table 3, working between 81 and 140 hours significantly predicted cigarette use for males (OR = 3.32, 95% CI = 1.53, 7.20).

Alcohol use among male participants was also significantly predicted by number of hours worked. Working between 41 and 80 hours in the past four weeks significantly predicted current alcohol use (OR = 1.89, 95% CI = 1.04, 3.43), as did working between 81 and 140 hours (OR = 2.54, 95% CI = 1.39, 4.65).

Table 3. Logistic regression analysis predicting males' current cigarette and alcohol use

Variable	N	B	S.E.	OR (95% CI)	R^2
Cigarettes					.058
1-40 hours	99				
41-80 hours	131	.35	.42	1.42 (.62, 3.22)	
81-140 hours	114	1.20**	.39	3.32 (1.53, 7.20)	
Alcohol					.039
1-40 hours	99				
41-80 hours	131	.64*	.30	1.89 (1.04, 3.43)	
81-140 hours	114	.93**	.31	2.54 (1.39, 4.65)	

Note: *p<.05, **p<.01.

For female participants, working 81 to 140 hours in the past four weeks significantly predicted current cigarette use (OR = 3.19, 95% CI = 1.59, 6.43) (see table 4).

Table 4. Logistic regression analysis predicting females' current cigarette and alcohol use

Variable	N	B	S.E.	OR (95% CI)	R^2
Cigarettes					.061
1-40 hours	97				
41-80 hours	107	.37	.38	1.45 (.69, 3.04)	
81-140 hours	100	1.16**	.36	3.19 (1.59, 6.43)	
Alcohol					.011
1-40 hours	97				
41-80 hours	107	-.11	.29	.90 (.51, 1.59)	
81-140 hours	100	.33	.29	1.39 (.78, 2.45)	

Discussion

As adolescents explore new roles and negotiate various developmental challenges, they increasingly begin to make independent decisions about their own behaviors in a variety of domains. Many young people decide to enter the workforce and obtain part-time employment outside the home. Historically, adolescent employment was more likely to consist of apprenticeship-type jobs, working on the family farm, or jobs taken to help a family through difficult economic times. However, in recent years, adolescents typically work part-time to make spending money for themselves or to save for their education rather than to develop a vocational identity or contribute to the family income. This is particularly true of the participants in this study, who worked in low-skilled, entry-level jobs.

In the present study, we found that young male and female employees working high-intensity hours (81 to 140 hours per month) were more than three times as likely to smoke cigarettes compared to those working low-intensity hours (40 hours or less per month). Furthermore, young men working moderate- or high-intensity hours were more likely to drink alcohol compared to young men working low-intensity hours, although work intensity was not associated with alcohol use for young women. These findings provide some evidence that employment in low-skill, entry-level jobs at certain levels of work intensity, is associated with an elevated risk of substance use for youth. Male employees may be at particular risk for alcohol use at moderate- and high-intensity work hours. The absence of a relationship between hours worked and substance use at low-intensity work hours suggests that the number of hours worked is a critical factor in understanding risk of substance use among adolescent and young adult employees.

Our findings are consistent with a growing number of studies that have found higher rates of cigarette use, alcohol use, illicit drug use, and heavy substance use among employed adolescents (1-3,5,15-16). These findings, along with other studies linking the intensity of hours worked to increases in risk for sexual risk-taking (17), delinquency (13) and psychological distress (4), suggest that youth who work are an important target population for preventive interventions. Research is needed to test the feasibility and effectiveness of preventive interventions targeting youth in the workplace.

A limitation of this study is the cross-sectional design, which limits the ability to examine the direction of effects. Work environments may increase substance use behavior through socialization processes such as workplace norms supporting use, greater availability of drugs from older coworkers, substance use at after-work social gatherings, and the availability of spending money that teens may decide to spend on drugs (18,19). On the other hand, adolescents with specific characteristics or behaviors may choose to work longer hours, and such selection processes may help explain the link between work intensity and substance use. Research has shown that students who receive poor grades or do not plan to attend college were more likely to want and attain higher levels of work intensity (14), and that disengagement from school often predicts entry into the workforce (13). Students who desire to work long hours have been found to engage in substance use and that this desire often exists prior to entering the workforce (14).

In contrast, some studies have found increases in problem behaviors among employed adolescents even after considering pre-employment differences. For example, Ramchand, Ialongo and Chilcoat (3) reported increased rates of tobacco use initiation among employed adolescents independent of selection effects. Steinberg, Fegley and Dornbusch (13) found support for both selection effects and socialization processes. These authors state that "In essence the operative process is not selection or socialization, but a dynamic and reciprocal process, in which adolescents both actively choose and are at the same time affected by the environments they encounter" (p. 178). Future research should use longitudinal designs to further examine the relative importance of selection and socialization processes in workplace substance use among youth.

In summary, the present findings suggest that for some adolescents, working more than half-time in low-skilled, entry-level positions can be associated with increased substance use risk. Workplace substance use prevention initiatives for this segment of the workforce may be warranted. More research is needed to help identify the work contexts, job types, and job responsibilites for which long work hours are related to adolescent substance use risk. This research would help to inform prevention efforts and assist policy makers in addressing this issue.

Acknowledgments

This research was supported by funds from the Substance Abuse and Mental Health Services Administration / Center for Substance Abuse Prevention (Grant 5UD1SP11134).

References

[1]   Safron DJ, Schulenberg JE, Bachman JG. Part-time work and hurried adolescence: The links among work intensity, social activities, health behaviors, and substance use. J Health Soc Behav 2001;42:425-49.

[2]   Valois RF, Dunham ACA, Jackson KL, Waller J. Association between employment and substance abuse behaviors among public high school adolescents. J Adolesc Health 1999;25:256-63.

[3]   Ramchand R, Ialongo NS, Chilcoat HD. The effect of working for pay on adolescent tobacco use. Am J Public Health 2007;97:2056-62.

[4]   Weller NF, Kelder SH, Cooper SP, Basen-Engquist K, Tortolero SR. School-year employment among high school students: Effects on academic, social, and physical functioning. Adolescence 2003;38:441-58.

[5]   McMorris BJ, Uggen C. Alcohol and employment in the transition to adulthood. J Health Soc Behavior 2000;41:276-94.

[6]   Mortimer JT, Finch MD, Ryu S, Shanahan M J, Call KT. The effects of work intensity on adolescent mental health, achievement and behavioral adjustment: New evidence from a prospective study. Child Dev 1996;67:1243-61.

[7]   Johnson MK. Further evidence on adolescent employment and substance use: Differences by race and ethnicity. J Health Soc Behav 2004;45:187-97.

[8]   Zimmer-Gembeck MJ, Mortimer JT. Adolescent work, vocational development, and education. Review of Educational Research 2006;76: 537-66.

[9]   Paternoster R, Bushway S, Brame R, Apel R. The effect of teenage employment on delinquency and problem behaviors. Social Forces 2003;82: 297-335.

[10]   Hansen DM, Jarvis PA. Adolescent employment and psychosocial outcomes: A comparison of two employment contexts. Youth and Society 2000;31:417-36.

[11]   Bellair PE, Roscigno VJ, McNulty TL. Linking local labor market opportunity to violent adolescent delinquency. J Res Crime Delinq 2003;40:6-33.

[12]   Kessler RC, Barber C, Beck A, Berglund P, Cleary PD, McKenas D, et al. The World Health Organization Health and Work Performance Questionnaire (HPQ). J Occup Environ Med 2003;45:156-74.

[13]   Steinberg L, Fegley S, Dornbusch SM. Negative impact of part-time work on adolescent adjustment: Evidence from a longitudinal study. Dev Psychol 1993;29:171-180.

[14]   Bachman JG, Safron DJ, Sy SR, Schulenberg JE. Wishing to work: New perspectives on how adolescents' part-time work intensity is linked to educational disengagement, substance use, and other problem behaviours. Int J Behav Development 2003;27:301-15.

[15]   Paschall MJ, Flewelling RL, Russell T. Why is work intensity associated with heavy alcohol use among adolescents? J Adolesc Health 2004;34:79-87.

[16]   Wu L, Schlenger W, Galvin D. (2003). The relationship between employment and substance use among students aged 12 to 17. J Adolesc Health 2003;32:5-15.

[17]   Valois RF, Dunham AC. Association between employment and sexual risk-taking behaviors among public high school adolescents. J Child Fam Stud 1998;7:147-59.

[18]   Arnett JJ. Adolescence and emerging adulthood: A cultural approach (2nd ed.). Upper Saddle River, NJ: Prentice Hall 2004.

[19]   Wright JP, Cullen FT, Williams N. The embeddedness of adolescent employment and participation in delinquency: A life course perspective. Western Criminol Rev 2002;4:1-19.

Section 4: Interventions

In: Public Health Concern   ISBN: 978-1-62948-424-2
Editors: J Merrick and A Tenenbaum   © 2014 Nova Science Publishers, Inc.

Chapter XIII

Decreasing student tendency towards smoking

Mohammad Ali Emamhadi, MD[*1],
Farideh Khodabandeh, MD[1], Maryam Jalilvand, MD[2],
Mina Hadian, MD[2] and Gholam Reza Heydari, MD[3]

[1]Legal Medicine Department, School of Medicine,
Shahid Beheshti University of Medical Sciences, Tehran
[2]Tohid Counseling and Psychological Services,
Ministry of Education and Training, Tehran
[3]Tobacco Prevention and Control Research Center, NRITLD,
Shahid Beheshti University of Medical Sciences, Tehran, Iran

Abstract

Smoking is the first cause of preventable morbidity and mortality in the world. In this chapter we compare different methods in reducing student tendency towards smoking. This semi-experimental study comprised all 7th grade students studying in middle schools throughout Iran. Students were divided into four groups: three study groups (social skills training, increasing knowledge and poster presentation) and one control group.

* Correspondence: Mohammad Ali Emamhadi, MD, Assistant professor of toxicology and forensic medicine, Department of Legal Medicine, NRITLD, Shahid Beheshti University, POBox 13185-1678, Tehran, Iran. E-mail: swt_f@yahoo.com.

Sampling method used was multi-phase cluster. The country was geographically divided into five districts (north, south, and east, west and central) and the provinces were selected randomly. A questionnaire was used to collect the data. These questionnaires were designed to evaluate the attitude and knowledge of students with regard to smoking and complications. Results: A total of 2,911 students with the mean age of 13 years were studied out of which 7.4% were smokers. There were significant differences between the study groups and the control group regarding the attitude and knowledge about the hazards of smoking and abuse of illicit substances. In other words, among the study groups, social skills training, building knowledge and poster presentation had the best results, respectively. In evaluating the preventive methods, social skills training group had the most negative attitude and the highest level of knowledge concerning the disadvantages and hazards of smoking and use of illegal substances. The greatest decrease in smoking was also observed in this group. Social skills training can be an effective preventive measure to control smoking by emphasizing self-respect, problem-solving skills and self restraint.

Introduction

Smoking is the first cause of preventable morbidity and mortality in the world. It is known as the most preventable cause of premature death (1). Studies have shown that, on average, smokers die nearly seven years earlier than non smokers (2). Smoking is responsible for 90% of lung cancers, 40% of other cancers, 50% of cardiovascular diseases, 75% of respiratory diseases, 12% of all deaths and 30% of deaths occurring in the age range of 30-50 years (3).

It seems that an epidemic of smoking and related morbidity and mortality is shifting towards developing countries (4). The most susceptible time for initiating tobacco use is during adolescence and early adulthood, between the ages 15 and 24 years (5). According to the Global Youth Tobacco Survey report; globally, 9.5% of adolescents currently use cigarettes (6), with nearly 25% of them trying their first cigarette before the age of 10 and 19.1% susceptible towards initiating smoking during the next year (7).

Some studies indicate that smoking is often associated with other forms of high-risk behaviors such as the early initiation of sexual intercourse, alcohol abuse and drug use (4). Considering the fact that the majority of Iran's population is youth, implementation of smoking control programs with special emphasis on preventing initiation among adolescents is a matter of importance. According to the first round of the GYTS in 2003, 2% of Iranian students (13-15 years) smoked cigarettes. This amount has increased to 3% in the second round of GYTS in 2007 (7).

We assessed different methods to know which of them is more effective in decreasing the tendency of students towards smoking.

Our sample

This was a Semi-Experimental study to evaluate the effect of preventive programs on changing the attitude and increasing the knowledge of students. Students were divided into four groups: Three study groups (social skills training, increasing knowledge and poster presentation) and 1 control group. The study population comprised all 7th grade students studying in middle schools throughout the country in the year 2005-2006. Sampling was done via multi-stage cluster. The country was geographically divided into 5 districts (north, south, and east, west as well as central). From districts, one province and from the south and west districts, two provinces were chosen randomly. Thus, Mazandaran, Khorasan, Tehran, Azerbaijan, Kurdistan, Kerman and Fars provinces were selected from the north, east, central, west and south areas respectively.

The capital of each province was selected out of which eight schools (four girl's schools and four boy's schools) were chosen. In each school, two classrooms with a mean number of 30 students were selected. Out of the selected 4 girl's or boy's schools, three study groups (1-Social skills training 2-Increasing the knowledge and 3- poster presentation) and one control group with no training were selected. All four groups filled out questionnaires. Then, the 1st and 2nd groups participated in a 6-session curriculum each lasting for 45 minutes (one session a week and a total of 1.5 months). During this time period, some posters related to the subjects were hung on the walls of the group three schools so that students could see them easily. Two months after completion of the course, all 4 groups filled out the questionnaires for the 2nd time.

Social skills training program was based on the social cognitive theory where the social influence increased competence. This program aims to teach some necessary skills to students to efficiently and effectively confront the social effects of smoking and in general, increases the students' ability to fight. It also promotes the mental health, primarily prevents the social damages of addiction and substance abuse, builds the knowledge, changes the attitude and behavior and eventually develops the ability for compatibility and compliance with the changing conditions of life through education and training personal and social skills.

Building the knowledge program was performed through holding an educational curriculum on complete and comprehensive description of short-term and long-term consequences of smoking. Questionnaires were designed to evaluate the attitude and knowledge of students towards smoking complications and smoking status of them and their family members. The reliability of knowledge and attitude questionnaires was estimated to be 0.87 and 0.89 respectively.

Descriptive statistical methods were used to analyze the data. To evaluate the correlations and differences, chi-square test, ANOVA and MANOVA were used with SPSS version 12 software.

Findings

A total of 2,911 students with the mean age of 13 years were studied out of which 49.5% were males and 50.5% were females; 23% were in the age range of 11-12 years, 75% were in the age range of 13-14 years and 2% were 15 years or older.

Thirty-one percent of students' family members were smokers out of which, in 27% of cases the father was smoker, in 1.8% the brother, in 0.5% the mother, in 0.3% the sister and in the remaining, more than 1 family member were smokers. Almost 7.4% of students (4.8% boys and 2.7% girls) were smokers. The difference between the two sexes was statistically significant (p=0.000). Also, there was a significant correlation between smoking by parents and cigarette smoking by their children (p<0.001).

Most of the smoker students initiated their smoking at the age of 11 (Table 1). The mean age of smoking initiation was 11.7 years. The prevalence of smoking was higher among Tehran and Mazandaran students. Smoking prevalence in Tehran was twice the total prevalence rate of the country.

The started reason for smoking was curiosity (5%), sadness (1.4%), offering by others (1.23%) and leisure (0.7%) (Table 2). Students acquired the cigarette mostly from others (4%) or buying it personally (3%). Boys bought cigarettes personally significantly more than girls (p=0.007). Students mostly smoked at a friend's house or at school (3%). Results demonstrated that boys mostly smoked at a friend's house or at school while girls mostly smoked at home (p=0.001).

Table 1. Age of smoking initiation

Age	10	11	12	13	14	15	total
Boy, n(%)	3(1.4%)	92(42.4%)	7(3.2%)	13(6%)	23(10.6%)	1(0.5%)	139(64.1%)
Girl, n(%)	1(0.5%)	51(23.5%)	9(4.1%)	7(3.2%)	10(4.6%)	----	78(35.9%)
Total	4(1.8%)	143(65.9%)	16(7.4%)	20(9.2%)	33(15.2%)	1(0.5%)	217(100%)

Table 2. The started reason for smoking

	Curiosity	Leisure	Sadness	Offering	Others	total
Boy, n(%)	79 (36.4%)	10(4.6%)	7(3.2%)	35(16.1%)	8(3.7%)	139(64.1%)
Girl, n(%)	50(23%)	4(1.8%)	8(3.7%)	12(5.5%)	4(1.8%)	78(35.9%)
Total	129(59.4%)	14(6.5%)	15(6.9%)	47(21.7%)	12(5.5%)	217(100%)

After our intervention, there was a significant difference between the study groups and the control group in terms of knowledge regarding smoking hazards and attitude towards smoking (p<0.001). In other words, social skills training showed the best result followed by knowledge increasing and poster presentation.

This was also true regarding decreasing or quitting smoking and the difference between the study groups and the control group was statistically significant in this regard (p<0.001, Table 3). Social skills training had the highest effect on quitting smoking followed by knowledge building and poster presentation.

Table 3. Scheffe Test data summary (mean difference of control group scores and study groups scores)

Variable	Group I	Group J	Mean difference between groups I and J	Standard deviation	P value
		Poster	-1.3789	0.18750	0.000
	Control	Knowledge	-2.1846	0.20570	0.000
		Life skills	-3.4000	0.19822	0.000
		Control	1.3789	0.18750	0.000
Decreased rate of smoking	Poster	Knowledge	-1.8057	0.19539	0.000
		Life skills	-2.0211	0.18750	0.000
		Control	2.1846	0.20570	0.000
	Knowledge	Poster	0.8057	0.19539	0.000
		Life skills	-1.2154	0.29570	0.000
		Control	3.4000	0.19522	0.000
	Life skills	Poster	2.0211	0.18750	0.000
		Knowledge	1.2154	0.20570	0.000

Discussion

The prevalence of smoking was 2.6% among girls and 4.8% among boys and 7.4% in general. A significant correlation was seen between gender and the rate of smoking. A national survey in the United States showed that 18% of 8th graders, 26% of 10th grade students and 23% of 12th grade students were daily smokers (8).

The difference between the results of our study and GYTS is probably caused by the different methods of smoking status evaluation. In our study all student who had the experience of smoking in the last 30 days were considered as smokers but in GYTS, the student who smoked daily were considered as smoker.

Out of 7.4% smoker students, almost half reported the age of smoking initiation to be less than age 12 with a mean of 11.3 years. Studies indicate that the age of smoking initiation has decreased (9) and some researchers reported that more than 95% of smokers have started smoking before the age of 19 (6).

In addition, this is why WHO in its 1999 report has emphasized that "if we could yield the tobacco use to zero in the 2nd decade of life, the overall rate in the society will decrease to less than 10% of the present rate" (10).

In this study, the main reason for smoking was curiosity followed by sadness and offering by the others. Students mostly acquired the cigarettes through others and smoked them usually at a friend's house.

This fact is in accord with other studies which show that juveniles usually imitate their peers in terms of law breaking behaviors such as smoking or substance abuse (11).

Evidence shows that attitudes affect the behavior (12-13). As it was shown in our study; there is a correlation between initiation of smoking and attitudes towards it (14-15).

Some experimental studies have confirmed the effect of attitude on substance use. For example, a study on 2646 students in 7th-grade of high school indicated that their current and anticipated smoking in the upcoming year was significantly correlated with their positive attitude towards smoking (16).

Our study results indicated that among the 3 afore-mentioned interventions, social skills training had the greatest impact on changing the attitude. Building a broader knowledge on health hazards of smoking was also effective in changing the attitude. However, in terms of knowledge, no difference was detected between the two methods of skill training and knowledge increasing. Our study results are in accord with previous studies (17-20).

In terms of increasing knowledge changing attitude and decreasing the rate of smoking, life skills training method was the best as it has been shown by other studies (18). Building knowledge was the same as life skills training in increasing the knowledge but it was less effective in changing the attitude and decreasing the rate of smoking. A poster was the least effective method in decreasing the students' tendency towards smoking.

Acknowledgment

The author would like to thank Farzan Institute for Research and Technology for technical assistance.

References

[1]   Centers for Disease Control and Prevention (1993) Smoking attributable mortality and years of potential life lost-United State, 1990. MMWR 1993;42:645-8.

[2]   Centers for Disease Control and Prevention. Cigarette Smoking- Related Mortality. Tobacco Information and prevention. Accessed 2013 Mar 20. URL: http://www.cdc.gov/tobacco/research_data/health_consequences/mortali.htm

[3]   US Department of Health and Human Services. The health consequences of smoking: A report of the surgeon general. Atlanta, GA: Centers for Disease Control and Prevention, 2004.

[4]   WHO. Report on the global tobacco epidemic. Geneva: WHO, 2008.

[5]   Baumrind D. Familial antecedents of adolescent drug use: a developmental perspective. NIDA Res Monogr 1985;56:13- 44.

[6]   Global Youth Tobacco Survey Collaborating Group. Differences in worldwide tobacco use by gender: finding from the Global Youth Tobacco Survey. J Sch Health 2003;73:207-15.

[7]   Trends in tobacco use among school students in the Eastern Mediterranean Region 2007. Accessed 2013 Mar 20. URL: http:// www. emro.who.int/tfi/ pdf/tobacco%20-among-school-students.pdf

[8]   Johnston LD, Malley PM, Bachman JG. National survey results on drug use from the monitoring the future study, 1975-2002, Vol. I: Secondary school students. Washington, DC: Department of Health and Human Services, 1995:46-54.

[9]   Monshi G, Samouie R, Valiani M. The effect of social skills training on preventing adolescent addiction in Isfahan. Abstract book of the 1st congress on scientific evaluation of law-breakers and preventive solutions, Khorasgan Islamic Azad University, 2005:109-93.

[10]   World Health Organization. A thorough review of global consumption, smoking rates, tobacco industry and national tobacco control action by country up to 1995-96. A global status report. Geneva: WHO 1999:62-9.

[11]   Bandura A. Self-efficacy: toward a unifying theory of behavioral change. Psychol Rev 1977;84:191- 215.

[12]   Botvin GJ, Griffin KW, Diaz T, Miller N, Ifill-Williams M. Smoking initiation and escalation in early adolescent girls: One − Year follow − up of a school - based prevention intervention for minority youth. J Am Med Womens Assoc 1999;54:139-43.

[13] Karimzadeh S. Efficacy of "disseminator of health" project in Kerman middle-school students in 2002-2003. Kerman education and training organization, unpublished paper, 2003:127-32

[14] Oetting ER, Beauvais F. Peer cluster theory, Socialization characteristics. and adolescent drug use: A path analysis. J Couns Psychol 1987;34:205-13.

[15] Eisen M, Zellman GL, Murray DM. Evaluating the lions- quest "skills for adolescence" drug education program. Second-year behavior outcomes. Addict Behav 2003;28:883- 97.

[16] Barkin SL, Smith KS, DuRant RH. Social skills and attitudes associated with substance use behaviors among young adolescents. J Adolesc Health 2002;30:448-54.

[17] Botvin GJ, Griffin KW, Paul E, Macaulay A. Preventing tobacco and alcohol use among elementary school students through Life Skills Training. J Child Adoles Subs Abuse 2003;12(4):121-7.

[18] Botvin GJ, Schinke SP, Epstein JA, Diaz T. Effectiveness of culturally- focused and generic skills training approaches to alcohol and drug abuse prevention among minority youths. Psychol Addict Behav 1994;8:116-27.

[19] Botvin GJ, Baker E, Botvin EM, Filazzola AD, Millman RB. Prevention of alcohol misuse through the development of personal and social competence: A pilot study. J Stud Alcohol 1984;45:550-2.

[20] Botvin GJ, Baker E, Dusenbury L, Tortu S, Botvin EM. Preventing adolescent drug abuse through a multimodal cognitive-behavioral approach: Results of a three-year study. J Consul Clin Psychol 1990;58:437-46.

In: Public Health Concern   ISBN: 978-1-62948-424-2
Editors: J Merrick and A Tenenbaum   © 2014 Nova Science Publishers, Inc.

Chapter XIV

Spiritually-based activities as a deterrent to substance abuse behavior

Patricia Coccoma, EdD, LCSW[*]
and Scott Anstadt, PhD, DCSW
Division of Social Work, College of Professional Studies,
Florida Gulf Coast University, Fort Myers,
Florida, US

In this chapter we explore an alternative health behavior to substance abuse behavior on a university campus, a constructed survey, the Substance Abuse Spirituality Discipline Survey (SASDS), which includes the DSES (1) and UNCOPE,(2) plus 7 researcher questions that query present use patterns and preference for potential campus programs, was used with a college population. The self administered web based survey was completed by 444 self selected college students over 18 years of age (F/M- 65%/35%) using StudentVoice, a web based platform. Internal consistency was high for both scales (DSES α= .96, SASDS α=.96) and acceptable for the UNCOPE α= .79. Items from each scale correlated (p< .01) indicated an inverse relationship: greater daily spiritual discipline, the lower endorsement of substance abuse criteria. Students who self describe binging behavior, using amount, frequency, and 'get high'

[*] Correspondence: Patricia Coccoma, EdD, LCSW, Florida Gulf Coast University, 10501 FGCU Blvd S AB3, 144 Fort Myers, FL 33965, United States. E-mail: pcoccoma@fgcu.edu.

criteria, were compared with non-binging students. Amount and 'get high' bingers showed an interest for spiritually focused behavior programs and less willingness to attend substance abuse treatment/education programs than non-bingers.

Introduction

Our study was conducted to explore the interest in an alternative health behavior to substance abuse behavior on a university campus, and therefore a constructed survey, the Substance Abuse Spirituality Discipline Survey (SASDS), which includes the DSES (1) and UNCOPE (2), plus 7 researcher questions was used. Campus life is not academics alone. Campus programs offered integrated academic pursuits while also offering the student information for a disciplined and balanced lifestyle. This study explores choices and interest in campus programming using a web based survey intended to identify any preferences of alternative behaviors of at risk substance abusing students. The study concentrated on one component of the socio-emotional dimension toward well being, the spiritual dimension, involving a health enhancing behavioral education and activities with moderate to no use of alcohol or other drugs.

Substance use on college campus is common and significant. Thirty four percent of college students screened had a harmful or harzardous drinking problem (3), while researched risk factors of substance abusers on campus found a misperception amongst those with a substance use disorder of others useage of alcohol and drugs than those who did not met the SUD criteria (4). A recent national survey (5) found that full time college students were more likely to use alcohol, binge drink or drink heavily than their peers who were not enrolled in college. The results also found that binge drinking amongst college students had increased more than 20% than when last surveyed in 2008.

Methods of prevention and intervention are diverse and efficient models in prevention regarding substance use and abuse on college and university campuses have been explored (6). These include the social influence model utilizing peer group influence, the life stress model where the college experience is viewed as a period of stress in one's life and lastly, the social ecology model where campus power holders shape and deliver policy in this area shaping the campus structure. The authors note that reducing physical vulnerabilities and environmental stressors while encouraging and exercising the practice of coping strategies, self esteem, and other resources appear to be consistent with a culture of prevention on a campus rather than an additive approach as the need arises (6). This knowledge encouraged the researchers to further explore which model(s) of prevention if any are utilized at a university campus in southeastern United States and to recommend potential preventive behavioral models to substance abuse on the campus.

Research noted campus substance use prevention programs best served college students when providing brief motivational interviewing and skill based training (7). The skill based training involved cognitive behavioral interventions such as "drink refusal skills" and consideration of negative consequence outcomes when using alcohol of substance. While reported that motivational interviewing, CBT and coping skills strategies showed merit in prevention and treatment of substance abuse (8).

Another study found a concentration of drinking levels in a rather small subgroup of students who tended to binge drink and denied obvious consequences of doing so (9) and others concluded using college survey data, reinforcers for convivial gatherings included perceived intimacy, privacy, and escape from stress through negative coping by use of the effects of the alcohol (10,11). It also concluded from the assessments, the two greatest sources of relief from perceived stressors fell within the realms of socio-emotional, and community support. Additional study found the use of alcohol for binge drinking college students as a 'feel high' coping mechanism against stress and depressive symptoms (12).

As is the case with many early stage users, a relatively small percentage identify as having a growing problem, let alone seek either educational or professional treatment services to address this issue (13). The substance abuse assessment and treatment literature point to change in use occurring when the user acknowledges an intrinsic motivation to reduce to eliminate use, thus moving from pre-contemplation to contemplation stage of change (14). One method of preventive intervention towards achieving this process is through the use of self administered survey methods heightening awareness of risks and alternatives to substance use (15). The self administration of a survey instrument by college students can be an experience of stress relief while at the same time providing a potential for personal awareness and educational information for self recognition (16).

To address questions of researchers' interest and hypothesis, a valid survey instrument based upon a set of professionally respected criteria would need to be chosen. At present, the research has not found such an instrument used with mainstream college students who may have a wide range of substance use patterns and who have not been identified by any external means as abusing substances.

The criterion for substance abuse is partially based upon one's use and/or abuse during any portion of the previous twelve months (17). If the individual is currently not abusing substances at the time of interview, perhaps even a majority of their past twelve months, they would continue to meet the criteria for substance abuse (17,18). Recent literature has encouraged the field to consider current behaviors and thus refine the operational definitions in DSM-V to be more aligned with current degrees of substance abuse manifested behavior (19). These include a threshold of 4-5 (F/M) drinks per setting at a frequency of at least once per week on average to reach binge drinking status; amount and frequency considerations (19).

Spirituality

Throughout the history of addiction treatment, spiritual emphases has remained a strong predictor of treatment success. Research viewed addiction as a spiritual problem that needed a spiritual answer (20) while the founders of AA made spirituality a cornerstone of their movement. Research reportedly has shown repeatedly that spiritual emphasis is often a reliable predictor of treatment success and recovery success when looked at from a retrospective perspective (21). However, until recently, religion and spirituality were often treated synonymously and most specifically within a Judeo-Christian framework (22-24).

Additionally, the professions recognition of one's spirituality or spiritual behaviors has had increasingly relevance in the treatment and preventive approaches in health and mental health including substance and alcohol use (25,26). The helping professionals have come to know that learning more of how individuals find meaning and balance (beliefs, behaviors) in their lives is of assistance when rendering interventions as needed. Spirituality has been defined in several ways. The researchers' interest of spirituality was focused upon activities that promote a meaning of a personal feeling of oneness and activities that support this feeling (27,28)

A spiritually based lifestyle with connection to deep or primary values may serve as an alternative to addictive behaviors and becomes ingrained through a daily spiritually based discipline. The nature of this discipline may vary but seems to have the validated elements (1); to help evoke and maintain generative changes leading to a turning away from self harming behavior, as well as turning towards behaviors leading to self efficacy, self-esteem, and futurity i.e. in the form of community (group membership), direct physical action (jogging, health promoting activities), or through enhanced efficacy and perception of control (29). Futurity is a term of faith of the Stages of Change model (14), when a significant amount of the progress from pre-contemplation to action stages is predicted by the degree to which the changer comes to continuously desire and strive after some future good whereby benefits of change are perceived as outweighing the costs of the change. All of these concepts imply a transformation of thinking and behavior whereby some more transcendent fundamental reality becomes most salient (30).

The Daily Spiritual Experiences scale (DSES) was derived through extensive exploration of the role of routine spiritual disciplines leading to health and well-being from a broad range of major world religious and spiritual traditions (1). Yet, the scale items do not seek the degree of adherence to religious dogma or practice. Rather, they look towards a very personal, transcendent and theistic experience as well as what is often thought of as unity with community and life in general (non-theistic). Using the DSES, the developers posit that the construct of daily spirituality in one's life may moderate perceived social stressors and promote health and well-being. In addition, a disposition of spiritual peace may reduce feelings of anxiety and depression, and may elevate mood and promote optimism and self esteem (1).

Studies have established relationships between daily spiritual experiences and positive mental health outcomes in samples of college students (31) who use spiritually based coping with specific stressors (e.g., bereavement, health problems, and family conflicts) as well as other personal difficulties (23,33,34).

Several studies (35,36) have found increases of daily experiences of spirituality, sense of purpose and meaning in life, and 'Self-Transcendence' as measured with the DSES were associated with reduction or absence of heavy drinking and psychological well being with various populations. However, these studies assume persons have by one means or another identified themselves as having a problem with alcohol or drugs and have sought treatment in a retroactive manner.

Efforts to bring substance abuse preventive programming to a university campus involving students, faculty and administrators, using a "top down" support of institutional change for such programming and adding to its effectiveness has been researched (37). The current study seeks to take an exploratory and preventive posture by crafting a scale which combines the elements of the DSES with the UNCOPE and in keeping with the social ecology model of prevention noted earlier would support a clear strategy for campus programming using outcome data gleaned directly from the student body. This would serve as exploratory study extending the current body of knowledge leading to a systematic procedure informing the planning of preventive interventions for reducing stress and ensuing risk of substance abuse on college and university campuses.

Our study

The use of a designed survey will reflect an association in university students either meeting or not meeting a substance abuse behavioral criteria as having a positive interest in a preventive spiritually based alternative behavior and activity program versus a traditional/educational or treatment based substance abuse program on a college/university campus. Our questions were:

- What is the relationship between routine spiritual behavioral practice and alcohol and drug use in a population of college students ?
- What is the preference of students who meet substance abuse behavioral criteria compared with non-abusing peers between spiritually based support groups and traditional educational/treatment prevention programs?

This study represents a targeted focus on random sample of the student population at of a southeast United States university. A notable number of the self selected students from this sample report behavior meeting the criteria for substance abuse (17) using an exploratory and formative process was used to investigate if spiritually based programming on campus may attract these students as an alternative to substance use activities

Upon approval of the university Institutional Review Board, the Office of Student Affairs was invited to assist with the selection process of student participants. Several meetings involving the researchers and the Vice President of Student Affairs occurred to determine how best to canvas student participants from the full campus. This collaborative effort resulted in the decision to utilize a web based internet site, hosted by Student Affairs, *Student Voice* (38). *Student Voice* is an educational platform that allows university and college campuses to receive ongoing support when conducting assessment initiatives. The researchers, in consultation with design coordinators at *Student*

Voice, established the format for the invitation of participants, completion of the consent form and the survey instruments. All students, 18 years and older, residing on or off campus were initially considered invited to complete the survey. A random sample of 4,000 college students resulted in a self selected response of 585 (14.6%), who were invited to complete the survey in *Student Voice*.

The authors utilized an electronic survey document, comprised of two validated surveys and seven additional questions designed by the authors, to gather data on intervention programs for campus use. This instrument was named the Substance Abuse and Spirituality Discipline Survey (SASDS).

The Daily Spiritual Experience Scale or DSES is designed to provide researchers a self-report measure of daily spiritual experiences. The sixteen-item scale includes constructs such as awe, gratitude, mercy, sense of connection with the transcendent, compassionate love, desire for closeness to God and includes measures of awareness of discernment/inspiration and transcendent sense of self. This measure has been increasingly used in the social sciences, and for examining changes in religious/spiritual experiences over time. It has been developed using a construct validation by deriving the spiritual meaning embedded in a variety of religious and cultural traditions, thus intended to be non-biased towards any particular religious faith and has a high internal reliability (a=.94 and .95)(1)

The UNCOPE is a six question screening instrument which taps into the DSM IV TR criteria for abuse/dependence with alcohol. It asks about excessive use, possible neglect of responsibilities, attempts to cut down, objections to use, preoccupation with use, and use to relieve emotional discomfort. The instrument has been validated across gender and ethnicity with a variety of populations. The internal consistency, or internal reliability, (a=.85) suggests that the items form a relatively homogenous index for dependence and that all of the items are indicative of dependence (2). Both instruments are found on the public domain and do not require consents for use.

The program based questions were designed by the researchers to gather data relevant for potential intervention or support programs on the university campus addressing education on spirituality and substance abuse, treatment for substance abuse, and spiritually oriented groups.

The present use patterns questions were modeled from the criteria of previous studies which suggest any of the three behavioral substance use patterns would constitute 'binge' use. These include an average binge amount per setting of 4 (female) to 5 (male) drinks, with at least a binge frequency of once per week, with the intention of getting high (19;39). Using these criteria, analysis would be carried out between groups that reported behavior meeting the threshold in frequency (FB), amount (AB), and desire to 'get high' (HB) and those that did not meet the frequency (NFB), amount (NAB), and desire to 'get high' (NHB).

Assurance to the confidentiality of respondents who completed survey instrument is of importance. The design for completing the survey was to initially electronically sign the consent form webpage. Once the consent form is completed, the survey opened to the participant. No identifying information was requested on the instrument. The outcome data from the survey instrument was securely stored in the web based assessment platform, Student Voice. The consent forms were filed within the Student Voice survey program in an electronically secure manner monitored by the Office of Student Affairs through the contracted use of the secure web based services of Student Voice. There were no paper counterparts to the electronic files. Electronically validated permissions issued through the Office of Student Affairs were limited to the authors and the *Student Voice* coordinator for support purposes only.

The survey did not need to be completed in one sitting allowing students to save and return to the instrument at their convenience. This option presents a potential limitation in that students would fail to return to a partially completed survey and thus limit the rate of return lessening the significance of the findings. To avoid this potential outcome, several random reminders were sent to all participants of the continued availability and closure date of the survey, encouraging additional participants and reminding current participants to complete any saved instruments.

Emails inviting representative sample of students enrolled at the university were sent by the Studentvoice staff with weekly reminders. The surveys were available for five weeks. Students were given a web link to the survey. The first page presented was the consent form approved by the university IRB. Only after the student clicked the box indicating agreement, would they transfer to the SASDS survey. Once the survey availability period was closed, the samples were tested for normality using SPSS. Demographics were also compared with general university population representative attributes.

Findings

A random selection of 4,000 college students was offered to participate in this study. Of those selected students, 585 (14.6%) responded initially to the survey. Of these, 444 students (11.1%) completed the survey. Of the 141 students who did not complete the survey, 80 (61%) did not proceed further after acknowledging the consent form and before responding to the first question on the survey. No explanations were provided by those students.

Descriptive demographics show that the sample who completed the survey differed from the larger cohort in age (X^2=53.949, DF=5, p<.000) with those answering the survey significantly younger and a higher representation of females (X^2= 25.316, DF=1, p<.000). The sample did represent the larger target population in ethnic background (X^2= 7.371, DF= 6, p< .288).

Internal consistency was established using Cronbach's Alpha which showed high reliability for both the UNCOPE (α =.80) and DSES (α=.96) when combined in SASDS for the survey sample participants.

Spearman's rho showed significant but weak negative correlations between scores on the DSES and UNCOPE (Spearman's rho= .217, p<.01). Further, the non-parametric Kruskal-Wallis test for ranks resulted in a significant difference between binge and non binge group median scores (X=18.602, df=6, p<.005), indicating that the higher DSES scores (higher spirituality), the lower substance abuse (higher UNCOPE scores) reflecting an inverse rank order effect (see table 1).

This study further analyzed self reported binge behavior regarding frequency of drinking episodes (FB group), amount per setting (AB group), and desire to get high (HB group), among student groups. Using a cut off of reported five drinks per setting and once per week or more of use, 30 (7.2%) of the students so imbibe, 101 (22.8%) of the respondents ingest more than five drinks at a setting, and 62 (13.9%) of the respondents use drugs to get high. The Kruskal-Wallis tests (Table1) indicate a significant difference between binge and non-binge groups' scores on spirituality (DSES scores) and on drug abuse (UNCOPE scores) variables through all three categories of binge behavior. FB, AB, and HB groups consistently scored lower values on DSES (lower spirituality) than non-binge groups (NFB, NAB, and NHB). Conversely, FB, AB, and HB groups consistently indicated higher substance abuse than NFB, NAB, and NHB groups reported through UNCOPE (see table 1).

Table 1. Kruskal-Wallis Mean Ranks on DSES and UNCOPE
by Binge Behavior

Binge Reported Behavior N= 444	DSES Mean Ranks	UNCOPE Mean Ranks
Frequency Binge: n= 47 Non-binge: n= 397	*Binge=180.6 Non-binge=227.45	*Binge=94.54 Non-binge=237.65
Amount Binge: n= 101 Non-binge: n= 343	***Binge= 171.55 Non-binge=237.50	***Binge=111.58 Non-binge=255.16
To 'Get High' Binge: n= 64 Non-binge: n= 380	***Binge=170.96 Non-binge=229.75	***Binge=81.13 Non-binge=244.40

Note:

* p < .01

*** p< .001

For DSES, where significance is noted, the smaller reading indicates less daily spiritual discipline.

For UNCOPE, where significance is noted, the smaller reading indicates more substance abuse behavior.

This exploratory research also included within the survey instrument, four questions (of the researchers 7 additional questions) concerning a preference and willingness to attend either an educational program on spirituality practices or substance abuse issues. The Kruskal-Wallis rank order test was used to determine student preferences in campus programming (substance abuse education versus treatment, and spirituality education by groups) among students who self reported binge behavior. Items in the SASDS inquired if students would attend educational programming or treatment services regarding drug use/abuse. Result found 60 students (13.5%) would attend educational programs and 47 students (10.6%) would attend treatment services. On the other hand, 186 (41.9%) of the respondents indicated they would prefer an educational program on spirituality. Results indicate a significant intention not to participate in treatment programs for substance abuse for FB, AB, and HB compared with NFB, NAB, and NHB.

No significant differences were found between these binge and non binge groups for substance abuse educational and spirituality programs if offered on campus (p> .05). The AB and HB significantly reported intention to enroll in spiritual activity group but not spiritual educational programs. Scores on both educational programs – substance abuse and spirituality, were not significantly different for all binge and non-binge groups. Results are presented in table 2.

Table 2. Kruskal-Wallis Mean Ranks on Attendance Preference for Substance Abuse or Spiritually Focused Programs

Binge Reported Behavior	Campus Program: Sub Abuse Education Mean Ranks	Campus Program: Sub Abuse Treatment Mean Ranks	Campus Program: Spirituality Education Mean Ranks	Campus Program: Spirituality Activity Groups Mean Ranks
Frequency Binge: n= 47 Non-binge: n= 397	*	****Binge=260.40 Non-binge=218.01	*	*
Amount Binge: n= 101 Non-binge: n= 343	*	****Binge=242.96 Non-binge=216.48	*	***Binge=195.44 Non-binge=230.47
To 'Get High' Binge: n= 64 Non-binge: n= 380	*	****Binge=248.40 Non-binge=217.11	*	**Binge=193.16 Non-binge=226.12

Note:
*P> .05
**p< .027
***p <.005
** **p< .001
The smaller of the two comparison groups in each cell indicates significantly greater willingness to attend the program described.

Discussion

The DSES and UNCOPE are reported to be validated instruments with a variety of population groups. Neither has been used specifically with a population of college students as in this study. Both scales maintained their internal consistency when compared with previous studies cited. The findings of this study, albeit limited to the current campus population sample, extend and support similar results in an adult sample (1).

The data in this study addresses the research question concerning the relationship between the scores of the UNCOPE and DSES of the population sample surveyed. While the relationship outcomes indicate strength, it was moderate to low, indicating scatter in the scores. To verify this finding additional replication study are recommended. Data suggest the more students practice a daily spiritual discipline, the less they are likely to abuse alcohol or drugs. Data also suggest those who lean more towards spiritually based activity groups use alcohol/drugs less frequently. Future qualitative research is recommended to explore the questions (i) whether college students who are substance abusing do so for emotional comfort? (ii)Do non substance abusing college students use spiritual activities for emotional comfort? (iii) Is there a relationship?

Additional instruments may be added for concurrent validation. Construction of the various scales used in an instrument may be a source of accuracy or of inaccuracy in obtaining valid results. Greater accuracy as to what is meant by the term 'spirituality', for example, might be achieved by using a larger variety of instruments which have more comprehensive content validity when used in concert.

While the survey did not need to be completed in one sitting this was considered a limitation prior to the commencement of the study. Several random reminders were sent to all participants of the continued availability and closure date of the survey, encouraging additional participants and reminding current participants to complete any saved instruments.

The researchers have worked closely with Student Affairs. Their awareness of the prevalent use of alcohol and drug use on campus has motivated that department to seek additional avenues of diversion from this practice. The data from the present use SASDS items (frequency, amount per setting and desire to get high) supports this awareness. The findings suggest an interest by abusing and non abusing students to make use of spiritual activity programs as a diversion from substance abusing behaviors. This information has been offered to the Student Affairs office.

The question becomes one of what types of campus programming may assist at risk students away from a culture of heavy and self destructive use of alcohol and drugs. Prevention efforts such as educational workshops and the infusion of information on substance abuse into the curriculum is one often used option. However, in this current study, only a small percentage of students who identify themselves as having binge consumption patterns also endorse attending campus substance abuse educational or treatment programs. Students who self reported binge drinking behavior (FB, AB, and HB) cluster ranked significantly low to the question of attendance in substance abuse treatment programs compared to their counterparts (NFB, NAB, and NHB), indicating possible denial of substance abuse and potential dependency.

On the other hand, a larger percentage of students showed an interest in attending an educational program and activity groups on spiritual behaviors and activities. Further, AB and HB students also cluster ranked high in willingness to attend these programs compared to their counterparts. This data provide an answer to this study's second research question concerning preference of services which specifically attracts these students, even more than their non-binging peers, to seek a spiritually oriented and disciplined focus.

Future research may provide additional analysis of the DSES results as obtained from a college student sample. For example, the construct of 'God' is used in many of the DSES items because qualitative interviews leading to the construction of the DSES revealed the concept of 'God' to be universally understood across religious orientations (1), perhaps providing a specific and inclusive format for spiritually based activity groups on campus. The DSES was not developed to address or support any specific religion. Further support for the use of this terminology is found in substance abuse recovery precedent. The founders of AA and other self help groups incorporated the concept of 'God' into their 12 Step programs and continue to use this reference to the present (40).

The addition of campus programming with spiritual behavior activity and educational component may attract those students who do or could become substance abusers. To the researcher's knowledge, this is the first study using an exploratory program evaluation model (web based survey) that explored potential resources that would intervene and/or divert from substance abusing behavior by students on a college campus. Further replication is indicated to expand and/or support these findings.

The survey was sent out to a random sample of one third of the total student population. The sample who responded was equivalent in ethnic diversity but at a younger age, and more heavily represented by females than the overall university demographical profile. This is considered a limitation of the study. The response rate on the survey was 14.6% is limiting. To achieve a greater response rate, a replication study is recommended. A working association with the Office of Student Affairs is also recommended to improve the response rate. It is also recommended that the investigators and Student Affairs address means of diminishing the numbers of surveys initiated but not completed.

The combined DSES and UNCOPE plus 7 researcher questions resulted in an instrument called Substance Abuse and Spirituality Discipline Survey (SASDS). The intent of this descriptive instrument is exploratory regarding substance abuse and spiritual activity program interventions.

The data suggested a negative association between abusive use of alcohol/drugs and the establishment of a daily spiritual behavioral discipline.

Students who self report binge behavior reject attendance in campus substance abuse treatment compared to their non-binging peers. Of noted interest to the researchers, students who self report binge drinking behavior in amounts typically consumed per setting and a desire to get high also endorsed attendance at spiritual activity groups, suggesting consideration of an additional avenue of diversion from substance use. More descriptive validation and replication is necessary to further explore how spiritual activities and spiritual educational groups on campus effect a reduction in substance abuse by college students.

References

[1]   Underwood G, Teresi J, The daily spiritual experience scale: development, theoretical description, reliability, exploratory factor analysis, and preliminary construct validity using health-related data. Ann Behav Med 2002;24:22-33.

[2]   Hoffman NG, Hunt DE, Rhodes WM, Riley K J UNCOPE: A brief substance ependence screen for use with arrestees. J Drug Issues 2003;33:19-44.

[3]   Wallerstein G, Pigeon S, Kopans B, Jacobs D, Aseltine R. Results of national alcohol screening day: college demographics, clinical characteristics, and comparison with online screening. J Am Coll Health 2007;55:6.

[4]   Cranford A, Eisenberg D, Serras AM. Substance use behaviors, mental health problems, and use of mental health services in a probability sample of college students. J Addict Behav 2009;34(2):134-145.

[5]   SAMSHA. National survey on drug use and health: volume I. summary of national findings. Accessed 2013 Jul 28. URL: http://oas.samhsa.gov/NSDUH/2k9NSDUH/2k9ResultsP.pdf

[6]   Coyne RK, Wagner DI, Hadley TD, Piles MA, Schorr-Owen V, Enderly MT. Applying primary prevention precepts to campus substance abuse programs. J Couns Dev 1994;72(6):603-7.

[7]   Larimer ME, Cronce JM. Identification, prevention and treatment: a review of individual-focused strategies to reduce problematic alcohol consumption by college students. J Stud Alcohol 2002;Suppl 14:148-63.

[8]   Larimer M, Kilmer J, Lee C. College student drug prevention: a review of individually – oriented prevention strategies. J Drug Issues 2004;35:2.

[9]   Caldwell PE. Drinking levels, related problems and readiness to change in a college sample. Alcohol Treat Q 2002;20(2):1-15.

[10]   O'Hare T. Measuring excessive alcohol use in college drinking contexts: the drinking context scale. Addict Behav 1997;22(4):469-77.

[11]   O'Hare T. Measuring problem drinking in first time offenders: Development and validation of the college alcohol problem scale (CAPS). J Subst Abuse Treat 1997;14(4):383-7.

[12] Harrell ZAT. Is gender relevant only for problem alcohol behaviors? An examination of correlates of alcohol use among college students. Addict Behav 2008;33:359-65.

[13] Caldeira KM. Kasperski SJ, Sharma E, Vincent KB, O'Grady KE, Wish ED, et al. College students rarely seek help despite serious substance use problems J Subst AbuseTreat 2009;37(4):368-78.

[14] Prochaska JO, Norcross JC, DiClemente CC. Changing for good. New York: William Morrow, 1994.

[15] Bersamin M, Paschall MJ, Fearnow-Kenney M, Wyrick D. Effectiveness of a web-based alcohol misuse and harm-prevention course among high and low risks students. J Am Coll Health 2007;55(4):247-54.

[16] Bersamin M, Paschall MJ, Fearnow-Kenney M, Wyrick D. Effectiveness of a web-based alcohol misuse and harm-prevention course among high and low risks students. J Am Coll Health 2007;55(4):247-54.

[17] American Psychiatric Association. Diagnostic and statistical manual of mental disorders, 4th ed. Washington, DC: Author, 1994

[18] Andersson G, Ghaderi A. Overview and analysis of the behaviourist criticism of the diagnostic and statistical manual of mental disorders (DSM). Clin Psychol 2006;10(2):67-77.

[19] Martin CS, Chung T, Kirisci L, Langenbucher JW. Item response theory analysis of diagnostic criteria for alcohol and cannabis use disorders in adolescents: implications for DSM-V. J Abnorm Psychol 2006;115(4):807-14.

[20] Zoja L. Cultural madness. In: Lopez-Pedraza R, ed. Cultural anxiety. New York Daimon, 1990:29-55.

[21] Koski-Jannes A, Turner N. Factors influencing recovery from different addictions. Addict Res 1999;7:469-92.

[22] Pargament KI. The psychology of religion and coping. New York: Guilford, 1997.

[23] Ellison CG, Levin JS. The religion-health connection: evidence, theory, and future directions Health Educ Behav 1998;25:700–20.

[24] Idler EL, Musick MA, Ellison CG, George LK, Krause N, Ory D, et al Measuring multiple dimensions of religion and spirituality for health research. Res Aging 2003;25:327–65.

[25] Hodge D, McGrew C. Spirituality, religion and the interrelationship: a nationally representative study. J Soc Work Educ 2006;42:3.

[26] Graf M. Counseling and spirituality: views from the profession. Rehabil Couns Bull 2008;51:4.

[27] Pava M. Spirituality in (and out) of the classroom: a pragmatic approach. J Bus Ethics 2007;73:3.

[28] Cook C. Addiction and spirituality. Addiction 2004;99(5):539-51.

[29] Fidler J. The holistic paradigm and general systems theory. In: Gray W, Fidler J, Battista J, eds. General systems theory and the psychological sciences. Seaside, CA: Intersystems Press, 1982.

[30] Gray RM. The Brooklyn program: innovative approaches to substance abuse treatment. Federal Probation Quart 2002;66(3):9-16.

[31] Ciarrocchi JW, Deneke E. Happiness and the varieties of religious experience: religious support, practices, and spirituality as predictors of well-being. In: Piedmont

RL, Moberg D, eds. Research in the social scientific study of religion. Boston, MA: Brill, 2004:211-33.

[32] Cook JA, Wimberly DW. If I should die before I wake: religious commitment and adjustment to death of a child. J Sci Study Relig 1983;22:222–38.

[33] Ellison CG, Boardman JD, Williams DR, Jackson JS. Stress, religious involvement, and mental health: findings from the 1995 Detroit area study. Soc Force 2001;80:215–49.

[34] Krause N. Exploring the stress-buffering effects of church-based and secular social support on self-rated health in late life. J Gerontol B Psychol Soc Sci 2006; 61:35–43.

[35] Robinson ER, Cranford JA, Webb JR, Brower KJ Six-month changes in spirituality, religiousness, and heavy drinking in a treatment-seeking sample. J Stud Alcohol Drugs 2007;68:282-91.

[36] Zenmore SE, Kaskutas LA. Helping, spirituality and alcoholics anonymous in recovery. J Stud Alcohol 2004;65(3):383-91.

[37] Burns CF, Consolvo CA. The development of a campus-based substance abuse prevention program. J Couns Dev 1992;70(5):639-41.

[38] Student Voice. Accessed 2013 Jul 28. URL: http://www.studentvoice.com/

[39] Cranford A, Eisenberg D, Serras AM. Substance use behaviors, mental health problems, and use of mental health services in a probability sample of college students. J Addict Behav 2009;34(2):134-45.

[40] Steigerwald F, Stone D. Cognitive restructuring and the 12-Step Program of Alcoholics Anonymous. J Subst Abuse Treat 1999;16(4):321-27.

In: Public Health Concern   ISBN: 978-1-62948-424-2
Editors: J Merrick and A Tenenbaum   © 2014 Nova Science Publishers, Inc.

Chapter XV

Buprenorphine for the management of opioid withdrawal

Emily Haesler, BN, PGradDipAdvNsg[*]
Australian Centre of Evidence Based Aged Care,
La Trobe University, Australia

This is a short summary report on the findings of a Cochrane review on the use of buprenorphine for the management of opioid withdrawal. Buprenorphine is effective for assisting users of opioid drugs to withdraw, through the its effect in reducing withdrawal signs and symptoms and promoting withdrawal completion. Buprenorphine may offer advantages over both methadone and alpha2-adrenergeic agonists for managing opioid withdrawal. Evidence suggests slower rather than fast buprenorphine dose tapering to be more effective.

Introduction

This systematic review sought to answer two questions:1)What is the effectiveness of buprenorphine compared to methadone, alpha2-adrenergeic

* Correspondence: Emily Haesler, BN, PGradDipAdvNsg, Research Fellow, School of Nursing and Midwifery, Deakin University. E-mail: hemily@deakin.edu.au.

agonists (clonidine or lofexidine) and oxazepam for the management of opioid withdrawal? and 2) What is the effectiveness of different doses and rates of buprenorphine dose reduction?

Relevance for nursing

Opioid dependency is a significant social and public health issue associated with increased mortality, transmission of disease and ongoing health care costs. Opioid withdrawal is characterised by acute and protracted phases. Signs and symptoms of acute withdrawal include irritability, anxiety, muscular pains, sweating, insomnia, and gastrointestinal symptoms. Acute withdrawal begins 6 to 48 hours after the last opioid dose, reaches peak intensity within 2 to 4 days and signs of physical withdrawal cease within 2 weeks. Protracted withdrawal, characterised by cravings, malaise and reduced well being, continues for up to 6 months.

Buprenorphine is a partial agonist that is to manage opioid withdrawal, both as an alternative maintenance therapy and for short periods to suppress signs and symptoms of acute withdrawal. The medication is proposed as an alternative to methadone, which is a drug of dependence in itself, and to clonidine, which is associated with significant adverse events.

Study characteristics

The review (1) included 22 studies (n=1736), of which 21 were randomised controlled trials (RCTs) and one was a quasi-randomised trial. Twelve of the included trials compared buprenorphine to clonidine or lofexidine, five trials compared buprenorphine to methadone, five trials compared different buprenorphine dose reduction rates, one trial compared different buprenorphine commencing doses and one trial compared buprenorphine to oxazepam. Some trials provided evidence for more than one comparison. In most trials randomisation and allocation concealment were unclear. Whilst half the of trials were at a high risk of bias for subjective outcomes, the vast majority had a low risk of bias for objective outcomes, addressed incomplete data and were free of selective reporting and other bias.

In 13 trials participants were withdrawing from heroin, in four trials participants had recently used heroin or methadone and in other trials the specific opioid of addiction was not described. Psychotropics, alcohol, benzodiazepine and cocaine dependency were often an exclusion criterion. In 13 of the trials, treatment was provided on an inpatient basis. Treatment regimens for buprenorphine varied. No attempt was made to calculate buprenorphine equivalent doses between trials due to the complexity required. Maximum doses ranged from 0.3mg to 3.6mg intramuscularly and from 1.2mg to 16mg as sublingual tablets. There was no age limit for the review or for most of the included trials, although one RCT was limited to adolescents. Mean ages ranged from 17 to 47 years and participants were predominantly male and came from a wide range of social backgrounds.

Outcome measures included withdrawal severity or various different scales, retainment in therapy until withdrawal complete, time spent in treatment, adverse events and indications of remaining opioid-free throughout treatment. Meta-analysis was undertaken where possible.

Summary of key evidence

The evidence from five trials comparing buprenorphine to methadone showed:

- no difference in withdrawal severity between treatments in 3 trials
- one trial showed significantly lower withdrawal severity for buprenorphine at days 8 and 14 of treatment.
- Two trials showed no difference in average length of time in treatment.
- Three trials reported no significant difference in adverse events.
- Four trials showed no significant difference in rates of treatment completion.

Evidence from 12 trials comparing buprenorphine to alpha2-adrenergeic agonists showed:

- Pooled results from 4 RCTs (n=432) showed a significantly lower peak withdrawal score for buprenorphine (p<0.001).
- Pooled results (2 trials, n=452) showed a significantly lower overall withdrawal score for buprenorphine (p<0.001). This was supported in an additional four trials that reported data in a format unsuitable for pooling.

- Significantly greater retainment time in treatment for buprenorphine (4 trials, p<0.001).
- Significantly higher rate of completion of treatment for buprenorphine with a number needed to treat of 4 (95% confidence intervals [CI] 6 to 3, p<0.001, 4 trials).
- Four trials reported higher rates of commencement of natrexone maintenance therapy following buprenorphine treatment for opioid withdrawal.
- There was no significant difference in number or frequency of adverse events.

Evidence from five trials comparing different rates of buprenorphine dose reduction showed:

- No significant difference in withdrawal intensity between dose reduction rates in 2 trials, one trial found greater withdrawal intensity for participants with a faster dose reduction rate, one trial reported increase in muscle aches in participants whose dose was reduced faster and the fifth trial found peak withdrawal was reached earlier when dose was reduced faster.
- Three trials reported either no significant adverse events, or no difference in rate of side effects between fast and slow dose reduction regimens.
- Data on likelihood of treatment completion was insufficient to reach conclusions.

One trial comparing buprenorphine to oxazepam reported significantly lower mean withdrawal intensity for buprenorphine, with no difference in completion of treatment or rate of adverse events.

One trial comparing high versus low buprenorphine doses reported no differences in suppression of withdrawal within the first 24 hours of treatment or time to discharge.

Best practice recommendations

Buprenorphine is as effective as tapered methadone dosage in managing opioid withdrawal; however, buprenorphine may offer advantages in a faster resolution of withdrawal symptoms and a slightly higher level of treatment completion.

Buprenorphine is more effective than alpha2-adrenergeic agonists for reducing withdrawal signs and symptoms, retaining patients in treatment and promoting treatment completion.Gradual tapering of buprenorphine doses may be more effective than fast dose titration.

References

[1] Gowing L, Ali R, White JM. Buprenorphine for the management of opioid withdrawal. Cochrane Database Syst Rev2009;3:CD002025.

In: Public Health Concern ISBN: 978-1-62948-424-2
Editors: J Merrick and A Tenenbaum © 2014 Nova Science Publishers, Inc.

Chapter XVI

Legislative smoking bans for reducing secondhand smoke exposure, smoking prevalence and tobacco consumption*

Janice Christie, PhD, MA, BSc, RN, RSCPHN
School of Nursing and Midwifery, Queen's University Belfast, Ireland

The purpose of this short communication is to discover the effectiveness of legislative tobacco smoking bans in reducing exposure to secondhand smoke, smoking prevalence and consumption. Tobacco smoking is the second most common cause of worldwide mortality and people subjected to secondhand smoke (i.e. passive smoking) are more like to succumb to disease. 'Smoking bans' aim to stop tobacco smoking within indoor public spaces and can be enacted through policies or legislation at national, state or community level. The health rationale for bans arises from biochemical and epidemiological evidence regarding the toxicity of smoking and association between smoking and secondhand smoking with disease. It is also assumed that any ban will be socially acceptable, involve minimal cost and achieve high compliance. Bans may positively affect the health of many people through reducing exposure to passive

* Correspondence: Professor Joav Merrick, MD, MMedSci, DMSc, Medical Director, Health Services, Division for Intellectual and Developmental Disabilities, Ministry of Social Affairs and Social Services, POBox 1260, IL-91012 Jerusalem, Israel. E-mail: jmerrick@zahav.net.il

smoking and giving environmental support for smokers who want to quit or reduce their consumption. Yet, the research evidence regarding the effect of bans required systematic review.

Introduction

A Cochrane systematic review was undertaken to answer the question: "does legislation to ban or restrict tobacco smoking reduce exposure to secondhand smoke and smoking behaviour?"

Relevance for nursing

Reduction in both tobacco smoking behaviours and secondhand smoke exposure through smoking bans impact the environment and hence population health. Often bans are accompanied by education regarding the need for the ban and support services to help smokers use the opportunity to quit or reduce their tobacco consumption. Therefore, smoking bans can be associated with the 'settings approach' to health promotion derived from the World Health Organisation's Ottawa Charter (1986). The Charter supports population and individual health through integrated, coordinated and multi-dimensional interventions.

Study characteristics

This was a Cochrane systematic review which produced a 'narrative only' synthesis of 50 studies. Studies (with reported numbers of participants) had samples of 24 to 10413 subjects. No randomised controlled trials were identified that met inclusion criteria, however, 13 non-randomised controlled studies (quasi-experimental designs) were included; of these 7 were located in a general or workplace setting and six related to hospital admissions. A further 37 studies which lacked a control group but recorded data from participants pre and post intervention were also included in the review. In 16 studies random sampling was used to select participants from a target population, 11 employed convenience sampling, five used randomly selected clusters of bars/public houses; five sampling methods were unclear and two studies utilised mixed sampling methods.

Included studies followed participants for a minimum of six months following the ban (except for eight studies which measured exposure to second hand smoke). No age, gender or geographic limit was placed on participants in reviewed studies. A risk of bias summary table was not included in the review however, the reviewers note that one source of bias (blinding participants to receipt of the intervention) was not possible in any study, due to the nature of the ban intervention. Eight study authors identified that small sample size or low statistical power was a limitation to the generalisability of their research findings.

The intervention was a legislative ban on indoor smoking that was comprehensive within indoor sites (a total ban in 40 studies). Ten studies were included which the reviewers classified as 'restrictions' as they allowed smoking in a designated area. Twenty two of the reviewed studies were conducted in workplaces, most of these (19) targeted the health of hospitality workers in bars or restaurants. The reviewed studies included bans in 13 countries, mostly the USA (17) or Scotland (8). Exposure to secondhand smoke was measured by participants' self-report recording either the 'duration of exposure' or 'percentage sample exposed'. Some researchers (mostly those conducting large population studies) also took biochemical samples e.g. saliva, to validate self-reports. Measures such as smoking prevalence, tobacco consumption, smoking cessation and respiratory, cardiac or sensory health outcomes were also considered by the reviewers.

Summary of key evidence

- All 31 studies that investigated secondhand smoking found this reduced following the introduction of a ban. The duration of smoke exposure reduced from 71% to 100% and the percentage of the population exposed to smoke reduced from 22% to 85%. Twelve studies recording Cotinine saliva biomarker levels in non-smokers, found reductions of 39% to 89%.

- In total, 23 studies reported a measure related to smoking behaviour. Of the 10 research reports which investigated smoking prevalence as a ban outcome measure (rather than an explanatory variable for secondhand smoke exposure), most found a small reduction in smoking and two studies identified no change. Thirteen reports that considered 'smoking consumption' (amount smoked) produced inconsistent evidence of change (except for the 8 of the 7 studies that recorded both prevalence and consumption, which reported reductions in both outcomes). In total, 7 studies looked at 'smoking cessation' (most in combination with smoking consumption); the two studies that only investigated smoking cessation found a modest decrease in the numbers of smokers.

- Ten out of 12 studies researching respiratory health reported a significant reduction in symptoms. All of the 10 papers recording sensory outcomes (e.g. effects on eyes, nose and throat), identified fewer symptoms. A reduction in hospital admissions for acute myocardial infarction was found in every study (10) investigating this outcome. In addition, 2 reports identified a reduction in death from acute coronary disease and one reported better prognosis for non-smoking acute coronary syndrome patients.

- Nine studies reported increased support for the ban following implementation and 2 reported no change. Most of the 9 papers that recorded compliance stated that there was good ban compliance; only one study reported that 31% of smokers had either not changed or increased smoking. Seven studies reported economic implications, of these 3 found no decrease in bar use, 2 no decrease in restaurant use and 2 identified a reduction in the purchase of cigarettes.

Best practice recommendations

- Smoking bans can positively affect health through reduction in secondhand smoke (in particular, for hospitality workers) and the need for hospital based cardiac care.
- Based on evidence from nine studies, compliance with bans is generally high and public support increases during the duration of the ban.
- Clinicians and policy makers may wish to consider how to support bans through individual and population based measures to enhance compliance and support people who wish to use the opportunity to stop smoking.

References

[1]   Callinan JE, Clarke A, Doherty K, Kelleher, C. Legislative smoking bans for reducing secondhand smoke exposure, smoking prevalence and tobacco consumption. Cochrane Database Syst Rev 2010;4:CD005992.

Section 5: Acknowledgments

In: Public Health Concern   ISBN: 978-1-62948-424-2
Editors: J Merrick and A Tenenbaum   © 2014 Nova Science Publishers, Inc.

Chapter XVII

About the editors

Joav Merrick, MD, MMedSci, DMSc, born in Copenhagen, is professor of pediatrics, child health and human development affiliated with Kentucky Children's Hospital, University of Kentucky, Lexington, Kentucky, United States and the Division of Pediatrics, Hadassah Hebrew University Medical Center, Mt Scopus Campus, Jerusalem, Israel, the medical director of the Health Services, Division for Intellectual and Developmental Disabilities, Ministry of Social Affairs and Social Services, Jerusalem, the founder and director of the National Institute of Child Health and Human Development in Israel. Numerous publications in the field of pediatrics, child health and human development, rehabilitation, intellectual disability, disability, health, welfare, abuse, advocacy, quality of life and prevention. Received the Peter Sabroe Child Award for outstanding work on behalf of Danish Children in 1985 and the International LEGO-Prize ("The Children's Nobel Prize") for an extraordinary contribution towards improvement in child welfare and well-being in 1987. E-mail: jmerrick@zahav.net.il

Ariel Tenenbaum, MD, born in Argentina, graduated from the Hadassah Hebrew University Medical School in Jerusalem, Israel. Specialist in pediatrics and the director of the Down Syndrome Center and the Intellectual and Developmental Disability Evaluation Center at the Department of Pediatrics, Hadassah Hebrew University Medical Centers, Mt Scopus Campus, Jerusalem. Medical consultant at the Feeding Disorder Clinic, and director of the Pre-adoption Evaluation Clinic. Several publications in the field of pediatric medicine, child and public health and especially around medical conditions in persons with Down syndrome. E-mail: tene@hadassah.org.il

In: Public Health Concern   ISBN: 978-1-62948-424-2
Editors: J Merrick and A Tenenbaum   © 2014 Nova Science Publishers, Inc.

Chapter XVIII

About the National Institute of Child Health and Human Development in Israel

The National Institute of Child Health and Human Development (NICHD) in Israel was established in 1998 as a virtual institute under the auspicies of the Medical Director, Ministry of Social Affairs and Social Services in order to function as the research arm for the Office of the Medical Director. In 1998 the National Council for Child Health and Pediatrics, Ministry of Health and in 1999 the Director General and Deputy Director General of the Ministry of Health endorsed the establishment of the NICHD.

Mission

The mission of a National Institute for Child Health and Human Development in Israel is to provide an academic focal point for the scholarly interdisciplinary study of child life, health, public health, welfare, disability, rehabilitation, intellectual disability and related aspects of human development. This mission includes research, teaching, clinical work, information and public service activities in the field of child health and human development.

Service and academic activities

Over the years many activities became focused in the south of Israel due to collaboration with various professionals at the Faculty of Health Sciences (FOHS) at the Ben Gurion University of the Negev (BGU). Since 2000 an affiliation with the Zusman Child Development Center at the Pediatric Division of Soroka University Medical Center has resulted in collaboration around the establishment of the Down Syndrome Clinic at that center. In 2002 a full course on "Disability" was established at the Recanati School for Allied Professions in the Community, FOHS, BGU and in 2005 collaboration was started with the Primary Care Unit of the faculty and disability became part of the master of public health course on "Children and society". In the academic year 2005-2006 a one semester course on "Aging with disability" was started as part of the master of science program in gerontology in our collaboration with the Center for Multidisciplinary Research in Aging. In 2010 collaborations with the Division of Pediatrics, Hadassah Hebrew University Medical Center, Jerusalem, Israel around the National Down Syndrome Center and teaching students and residents about intellectual and developmental disabilities as part of their training at this campus.

Research activities

The affiliated staff have over the years published work from projects and research activities in this national and international collaboration. In the year 2000 the International Journal of Adolescent Medicine and Health and in 2005 the International Journal on Disability and Human Development of De Gruyter Publishing House (Berlin and New York) were affiliated with the National Institute of Child Health and Human Development. From 2008 also the International Journal of Child Health and Human Development (Nova Science, New York), the International Journal of Child and Adolescent Health (Nova Science) and the Journal of Pain Management (Nova Science) affiliated and from 2009 the International Public Health Journal (Nova Science) and Journal of Alternative Medicine Research (Nova Science). All peer-reviewed international journals.

National collaborations

Nationally the NICHD works in collaboration with the Faculty of Health Sciences, Ben Gurion University of the Negev; Department of Physical Therapy, Sackler School of Medicine, Tel Aviv University; Autism Center, Assaf HaRofeh Medical Center; National Rett and PKU Centers at Chaim Sheba Medical Center, Tel HaShomer; Department of Physiotherapy, Haifa University; Department of Education, Bar Ilan University, Ramat Gan, Faculty of Social Sciences and Health Sciences; College of Judea and Samaria in Ariel and in 2011 affiliation with Center for Pediatric Chronic Diseases and National Center for Down Syndrome, Department of Pediatrics, Hadassah Hebrew University Medical Center, Mount Scopus Campus, Jerusalem.

International collaborations

Internationally with the Department of Disability and Human Development, College of Applied Health Sciences, University of Illinois at Chicago; Strong Center for Developmental Disabilities, Golisano Children's Hospital at Strong, University of Rochester School of Medicine and Dentistry, New York; Centre on Intellectual Disabilities, University of Albany, New York; Centre for Chronic Disease Prevention and Control, Health Canada, Ottawa; Chandler Medical Center and Children's Hospital, Kentucky Children's Hospital, Section of Adolescent Medicine, University of Kentucky, Lexington; Chronic Disease Prevention and Control Research Center, Baylor College of Medicine, Houston, Texas; Division of Neuroscience, Department of Psychiatry, Columbia University, New York; Institute for the Study of Disadvantage and Disability, Atlanta; Center for Autism and Related Disorders, Department Psychiatry, Children's Hospital Boston, Boston; Department of Paediatrics, Child Health and Adolescent Medicine, Children's Hospital at Westmead, Westmead, Australia; International Centre for the Study of Occupational and Mental Health, Düsseldorf, Germany; Centre for Advanced Studies in Nursing, Department of General Practice and Primary Care, University of Aberdeen, Aberdeen, United Kingdom; Quality of Life Research Center, Copenhagen, Denmark; Nordic School of Public Health, Gottenburg, Sweden; Scandinavian Institute of Quality of Working Life, Oslo, Norway; The Department of Applied Social Sciences (APSS) of The Hong Kong Polytechnic University Hong Kong.

Targets

Our focus is on research, international collaborations, clinical work, teaching and policy in health, disability and human development and to establish the NICHD as a permanent institute at one of the residential care centers for persons with intellectual disability in Israel in order to conduct model research and together with the four university schools of public health/medicine in Israel establish a national master and doctoral program in disability and human development at the institute to secure the next generation of professionals working in this often non-prestigious/low-status field of work.

Contact

Joav Merrick, MD, MMedSci, DMSc
Professor of Pediatrics, Child Health and Human Development
Medical Director, Health Services, Division for Intellectual and Developmental Disabilities, Ministry of Social Affairs and Social Services, POB 1260, IL-91012 Jerusalem, Israel.
E-mail: jmerrick@zahav.net.il

In: Public Health Concern   ISBN: 978-1-62948-424-2
Editors: J Merrick and A Tenenbaum   © 2014 Nova Science Publishers, Inc.

Chapter XIX

About the book series "Health and Human Development"

Health and human development is a book series with publications from a multidisciplinary group of researchers, practitioners and clinicians for an international professional forum interested in the broad spectrum of health and human development. Books already published:

- Merrick J, Omar HA, eds. Adolescent behavior research. International perspectives. New York: Nova Science, 2007.
- Kratky KW. Complementary medicine systems: Comparison and integration. New York: Nova Science, 2008.
- Schofield P, Merrick J, eds. Pain in children and youth. New York: Nova Science, 2009.
- Greydanus DE, Patel DR, Pratt HD, Calles Jr JL, eds. Behavioral pediatrics, 3 ed. New York: Nova Science, 2009.
- Ventegodt S, Merrick J, eds. Meaningful work: Research in quality of working life. New York: Nova Science, 2009.
- Omar HA, Greydanus DE, Patel DR, Merrick J, eds. Obesity and adolescence. A public health concern. New York: Nova Science, 2009.
- Lieberman A, Merrick J, eds. Poverty and children. A public health concern. New York: Nova Science, 2009.

- Goodbread J. Living on the edge. The mythical, spiritual and philosophical roots of social marginality. New York: Nova Science, 2009.
- Bennett DL, Towns S, Elliot E, Merrick J, eds. Challenges in adolescent health: An Australian perspective. New York: Nova Science, 2009.
- Schofield P, Merrick J, eds. Children and pain. New York: Nova Science, 2009.
- Sher L, Kandel I, Merrick J, eds. Alcohol-related cognitive disorders: Research and clinical perspectives. New York: Nova Science, 2009.
- Anyanwu EC. Advances in environmental health effects of toxigenic mold and mycotoxins. New York: Nova Science, 2009.
- Bell E, Merrick J, eds. Rural child health. International aspects. New York: Nova Science, 2009.
- Dubowitz H, Merrick J, eds. International aspects of child abuse and neglect. New York: Nova Science, 2010.
- Shahtahmasebi S, Berridge D. Conceptualizing behavior: A practical guide to data analysis. New York: Nova Science, 2010.
- Wernik U. Chance action and therapy. The playful way of changing. New York: Nova Science, 2010.
- Omar HA, Greydanus DE, Patel DR, Merrick J, eds. Adolescence and chronic illness. A public health concern. New York: Nova Science, 2010.
- Patel DR, Greydanus DE, Omar HA, Merrick J, eds. Adolescence and sports. New York: Nova Science, 2010.
- Shek DTL, Ma HK, Merrick J, eds. Positive youth development: Evaluation and future directions in a Chinese context. New York: Nova Science, 2010.
- Shek DTL, Ma HK, Merrick J, eds. Positive youth development: Implementation of a youth program in a Chinese context. New York: Nova Science, 2010.
- Omar HA, Greydanus DE, Tsitsika AK, Patel DR, Merrick J, eds. Pediatric and adolescent sexuality and gynecology: Principles for the primary care clinician. New York: Nova Science, 2010.
- Chow E, Merrick J, eds. Advanced cancer. Pain and quality of life. New York: Nova Science, 2010.

- Latzer Y, Merrick, J, Stein D, eds. Understanding eating disorders. Integrating culture, psychology and biology. New York: Nova Science, 2010.
- Sahgal A, Chow E, Merrick J, eds. Bone and brain metastases: Advances in research and treatment. New York: Nova Science, 2010.
- Postolache TT, Merrick J, eds. Environment, mood disorders and suicide. New York: Nova Science, 2010.
- Maharajh HD, Merrick J, eds. Social and cultural psychiatry experience from the Caribbean Region. New York: Nova Science, 2010.
- Mirsky J. Narratives and meanings of migration. New York: Nova Science, 2010.
- Harvey PW. Self-management and the health care consumer. New York: Nova Science, 2011.
- Ventegodt S, Merrick J. Sexology from a holistic point of view. New York: Nova Science, 2011.
- Ventegodt S, Merrick J. Principles of holistic psychiatry: A textbook on holistic medicine for mental disorders. New York: Nova Science, 2011.
- Greydanus DE, Calles Jr JL, Patel DR, Nazeer A, Merrick J, eds. Clinical aspects of psychopharmacology in childhood and adolescence. New York: Nova Science, 2011.
- Bell E, Seidel BM, Merrick J, eds. Climate change and rural child health. New York: Nova Science, 2011.
- Bell E, Zimitat C, Merrick J, eds. Rural medical education: Practical strategies. New York: Nova Science, 2011.
- Latzer Y, Tzischinsky. The dance of sleeping and eating among adolescents: Normal and pathological perspectives. New York: Nova Science, 2011.
- Deshmukh VD. The astonishing brain and holistic consciousness: Neuroscience and Vedanta perspectives. New York: Nova Science, 2011.
- Bell E, Westert GP, Merrick J, eds. Translational research for primary healthcare. New York: Nova Science, 2011.
- Shek DTL, Sun RCF, Merrick J, eds. Drug abuse in Hong Kong: Development and evaluation of a prevention program. New York: Nova Science, 2011.

- Ventegodt S, Hermansen TD, Merrick J. Human Development: Biology from a holistic point of view. New York: Nova Science, 2011.
- Ventegodt S, Merrick J. Our search for meaning in life. New York: Nova Science, 2011.
- Caron RM, Merrick J, eds. Building community capacity: Minority and immigrant populations. New York: Nova Science, 2012.
- Klein H, Merrick J, eds. Human immunodeficiency virus (HIV) research: Social science aspects. New York: Nova Science, 2012.
- Lutzker JR, Merrick J, eds. Applied public health: Examining multifaceted Social or ecological problems and child maltreatment. New York: Nova Science, 2012.
- Chemtob D, Merrick J, eds. AIDS and tuberculosis: Public health aspects. New York: Nova Science, 2012.
- Ventegodt S, Merrick J. Textbook on evidence-based holistic mind-body medicine: Basic principles of healing in traditional Hippocratic medicine. New York: Nova Science, 2012.
- Ventegodt S, Merrick J. Textbook on evidence-based holistic mind-body medicine: Holistic practice of traditional Hippocratic medicine. New York: Nova Science, 2012.
- Ventegodt S, Merrick J. Textbook on evidence-based holistic mind-body medicine: Healing the mind in traditional Hippocratic medicine. New York: Nova Science, 2012.
- Ventegodt S, Merrick J. Textbook on evidence-based holistic mind-body medicine: Sexology and traditional Hippocratic medicine. New York: Nova Science, 2012.
- Ventegodt S, Merrick J. Textbook on evidence-based holistic mind-body medicine: Research, philosophy, economy and politics of traditional Hippocratic medicine. New York: Nova Science, 2012.
- Caron RM, Merrick J, eds. Building community capacity: Skills and principles. New York: Nova Science, 2012.
- Lemal M, Merrick J, eds. Health risk communication. New York: Nova Science, 2012.
- Ventegodt S, Merrick J. Textbook on evidence-based holistic mind-body medicine: Basic philosophy and ethics of traditional Hippocratic medicine. New York: Nova Science, 2013.
- Caron RM, Merrick J, eds. Building community capacity: Case examples from around the world. New York: Nova Science, 2013.

- Steele RE. Managed care in a public setting. New York: Nova Science, 2013.
- Srabstein JC, Merrick J, eds. Bullying: A public health concern. New York: Nova Science, 2013.
- Thavarajah N, Pulenzas N, Lechner B, Chow E, Merrick J, eds. Advanced cancer: Managing symptoms and quality of life. New York: Nova Science, 2013.
- Stein D, Latzer Y, eds. Treatment and recovery of eating disorders. New York: Nova Science, 2013.
- Sun J, Buys N, Merrick J. Health promotion: Community singing as a vehicle to promote health. New York: Nova Science, 2013.
- Pulenzas N, Lechner B, Thavarajah N, Chow E, Merrick J, eds. Advanced cancer: Managing symptoms and quality of life. New York: Nova Science, 2013.
- Sun J, Buys N, Merrick J. Health promotion: Strengthening positive health and preventing disease. New York: Nova Science, 2013.
- Merrick J, Israeli S, eds. Food, nutrition and eating behavior. New York: Nova Science, 2013.

Contact

Professor Joav Merrick, MD, MMedSci, DMSc
Medical Director, Health Services, Division for Intellectual and Developmental Disabilities, Ministry of Social Affairs and Social Services
POBox 1260, IL-91012 Jerusalem, Israel
E-mail: jmerrick@zahav.net.il

Section 6: Index

Index

A

abuse, 95, 138, 141, 143, 164, 173, 174, 176, 178, 179, 182, 183, 184, 186, 203, 211
academic performance, 88, 93
access, 11, 64, 106, 151, 152
accessibility, 129
acclimatization, 70
adaptation, 50, 70
adjustment, 36, 142, 158, 188
administrators, 177
adolescent addiction, 170
adolescent adjustment, 158
adolescent behavior, 34, 36, 37, 40, 41, 42, 43, 49, 50
adolescent boys, 92
adolescent drinking, 36, 51, 115, 120
adolescent problem behavior, 35, 36, 151
adulthood, 17, 54, 83, 129, 159, 164
adults, 3, 4, 10, 18, 22, 54, 62, 65, 67, 82, 92, 93, 95, 108, 127
adverse effects, 68, 81, 146
adverse event, 190, 191, 192
advertisements, 5, 9, 13, 15, 16, 17, 90
advocacy, 203
African Americans, 17, 51
agonist, 190
AIDS, 212
airways, 78

Alaska, 152
alcohol abuse, 138, 139, 145, 164
alcohol consumption, 81, 113, 114, 115, 116, 117, 118, 120, 121, 122, 124, 125, 126, 127, 128, 130, 140, 146, 186
alcohol use, 34, 35, 37, 38, 41, 42, 43, 49, 50, 52, 78, 114, 116, 127, 128, 129, 139, 147, 150, 151, 153, 154, 155, 156, 158, 171, 176, 186
alcohol withdrawal, 141, 142, 146
alcoholics, 140, 143, 188
alcoholism, 65
alternative behaviors, 174
alters, 79
American Psychiatric Association, 187
amygdala, 81
anemia, 130
anger, 63, 90
ANOVA, 166
antisocial behavior, 114, 151
anxiety, 4, 63, 68, 82, 141, 177, 187, 190
appetite, 4, 55, 78
Argentina, 203
arousal, 69
arthritis, 143
Asia, 31, 108
Asian countries, 105
assessment, 38, 117, 175, 178, 180
assets, 117
asthma, 10, 17, 18, 23, 143

atmosphere, 106, 140
attachment, 114
attitudes, 9, 11, 15, 16, 29, 63, 65, 85, 87,
    90, 95, 98, 99, 100, 101, 103, 104, 106,
    108, 169, 171
authorities, 50, 87, 139, 143
autonomy, 40
awareness, 24, 35, 63, 106, 175, 179, 184
Azerbaijan, 165

B

Bahrain, 87, 92, 95
ban, 195, 196, 197, 198
base, 87
beer, 118
behavioral change, 170
behavioral models, 175
behavioral problems, 117
behaviors, 4, 18, 19, 28, 34, 35, 36, 37, 39,
    42, 43, 48, 49, 50, 51, 54, 64, 66, 108,
    114, 116, 125, 127, 129, 130, 150, 156,
    157, 158, 164, 169, 171, 176, 184, 186,
    188
benefits, 22, 54, 58, 60, 79, 132, 137, 150,
    177
benzodiazepine, 191
beverages, 118, 119, 122, 139
bias, 16, 107, 127, 190, 197
binge drinking, 174, 175, 176, 184, 186
bipolar disorder, 81
birds, 127
birth weight, 23
births, 23
bivariate analysis, 116
blame, 74, 80, 144
blood, 138, 143
blood stream, 138
BMA, 37
body mass index (BMI), 78, 83
bonding, 36
bonds, 36, 114
boredom, 62, 93
brain, 82, 83, 138, 211
breathing, 4, 78

breathlessness, 93
brothers, 93
businesses, 151

C

caffeine, 22
calcium, 22
caloric intake, 78, 83
cancer, 4, 23, 29, 30, 210, 213
cannabis, 16, 138, 139, 143, 145, 146, 147,
    187
carbon, 16, 94
carbon monoxide, 16, 94
carcinogenicity, 23
carcinoma, 27, 31
cardiovascular disease, 164
caregivers, 38
Caribbean, 18, 139, 142, 143, 145, 146, 211
Caribbean countries, 139
causal inference, 64, 127
causality, 19, 64, 81
causation, 64
CDC, 11, 17, 94
challenges, 3, 82, 156
chaos, 138
chemicals, 30
Chicago, 108, 207
child abuse, 210
child development, 130
child maltreatment, 212
childhood, 54, 140, 211
children, 19, 22, 35, 36, 62, 72, 88, 95, 126,
    127, 128, 130, 138, 143, 166, 209
Chile, 95, 113, 116, 117, 128, 129
China, 105, 106, 108
chronic illness, 210
cigarette smoke, 10, 16, 86
cigarette smokers, 10, 16, 86
cigarette smoking, 5, 9, 10, 11, 14, 15, 17,
    22, 23, 24, 27, 28, 65, 85, 86, 87, 90, 92,
    93, 94, 95, 105, 108, 109, 150, 166
cities, 34, 56, 63, 108
city, 29, 85, 94, 95
civil servants, 83

classes, 11, 28, 87, 88, 100, 140
classification, 54, 62, 63, 65, 88, 95, 130, 153
classroom, 58, 187
classroom teacher, 58
clinical disorders, 82
clinical presentation, 147
clinical trials, 132
closure, 180, 184
cluster theory, 171
clustering, 10, 18
clusters, 196
cocaine, 144, 145, 191
coercion, 40
cognition, 83
cognitive function, 71, 83
cognitive theory, 165
cognitive-behavioral therapy, 80
collaboration, 37, 38, 206, 207
collage, 31
college campuses, 178
college students, 65, 100, 108, 173, 174, 175, 177, 178, 179, 181, 183, 184, 186, 188
colleges, 21, 24, 25, 28
color, 4
commercial, 4, 132
communication, 34, 36, 37, 39, 42, 43, 44, 51, 129, 195, 212
community, 23, 24, 28, 29, 30, 31, 34, 35, 36, 65, 98, 99, 105, 132, 175, 177, 195, 212
community support, 175
comparative analysis, 83
compatibility, 165
competing interests, 17
complexity, 191
compliance, 24, 165, 195, 198
complications, 164, 166
compulsive behavior, 81
computer, 37
computing, 115
conditioning, 80
confidentiality, 24, 94, 180
conflict, 9, 141, 151

conflict of interest, 141
conflict resolution, 151
congress, 170
connectivity, 82
consciousness, 211
consensus, 153
consent, 25, 57, 152, 179, 180, 181
construct validity, 186
construction, 185
consumers, 118
consumption, xiii, 22, 24, 30, 31, 115, 116, 120, 121, 122, 125, 126, 127, 128, 139, 140, 143, 145, 170, 184, 195, 196, 197, 198, 199
consumption patterns, 30, 184
control group, 163, 165, 168, 196
control measures, 98, 99, 105
controlled studies, 196
controlled trials, 190, 196
controversial, 132, 140
cooking, 40
coping strategies, 67, 71, 73, 75, 80, 175
correlation(s), 42, 43, 51, 59, 73, 75, 166, 168, 169, 181
correlation analysis, 75
cosmetic, 4
cost, 82, 93, 134, 195
Costa Rica, 139, 140, 141, 143, 144
cotinine, 16
cough, 86, 93
counseling, 50
craving, 86, 88
creative abilities, 139
creativity, xii, 137, 138, 139, 143
criminal violence, 139, 147
criticism, 117, 119, 187
crop, 108
cross sectional study, 21, 24, 29
cross-cultural differences, 69
cross-sectional study, 11, 19, 107
cultivation, 98
cultural influence, 34, 49
cultural tradition, 179
cultural values, 34

culture, 22, 34, 36, 40, 49, 50, 64, 105, 145, 175, 184, 211
curriculum, 11, 165, 166, 184

D

damages, 165
dance, 211
data analysis, 17, 210
data collection, 37, 72, 98, 133, 152
data set, 72, 101, 105, 117, 118
deaths, 23, 164
deficiency, 78, 130
deficit, 4, 5
delinquency, 33, 34, 35, 37, 38, 41, 42, 43, 44, 45, 46, 47, 49, 51, 129, 151, 156, 158, 159
delinquent behavior, 35, 41, 50, 51
demographic characteristics, 88, 89
demographic data, 26, 72
demographic factors, 43
denial, 184
Denmark, 207
Department of Education, 207
Department of Health and Human Services, 170
dependent variable, 60, 116, 118, 154
depression, 4, 5, 63, 80, 81, 140, 141, 145, 177
depressive symptoms, 109, 175
deprivation, 29, 82
depth, 105
developed countries, 17, 86, 98
developing countries, 86, 93, 117, 139, 164
deviation, 168
diabetes, 23
diagnostic criteria, 187
disability, xii, 137, 146, 203, 205, 206, 208
discomfort, 179
discriminant analysis, 73
diseases, 5, 18, 23, 164
disorder, 4, 5, 174
disposition, 177
dissatisfaction, 70, 82
distress, 68, 70, 71, 75, 80, 81, 82

distribution, 54, 70, 72, 93, 118
Dominican Republic, 129
dosage, 144, 192
Down syndrome, 203
dreaming, 73, 74, 80
drinking pattern(s), 114, 115, 128
drug abuse, 28, 142, 171, 181
drug addict, 146
drug education, 171
drug testing, 131, 132, 133, 134
drugs, 10, 17, 35, 36, 38, 41, 74, 86, 129, 132, 139, 144, 147, 153, 156, 174, 177, 181, 183, 184, 185, 189
DSM, 179
dyslipidemia, 23

E

eating disorders, 211, 213
ecology, 174, 178
economic status, 106, 108
editors, 203
educated women, 105
education, 28, 66, 79, 90, 107, 117, 152, 153, 156, 158, 166, 171, 174, 179, 182, 196, 211
educational attainment, 88
educational programs, 182
elders, 40
elementary school, 171
emotion, 83
emotional disorder, 81
emotional stimuli, 81
emphysema, 4
employees, xii, 149, 151, 152, 156
employers, 132, 151
employment, 132, 150, 151, 152, 156, 157, 158, 159
employment status, 152
encouragement, 115, 129
energy, 70, 78, 83
energy expenditure, 79, 83
engineering, 22, 24
enrollment, 11
environment, 15, 68, 106, 107, 142, 196

environmental factors, 4, 106, 107
environmental stress, 174
environmental tobacco, 11
environments, 156, 157
enzymes, 86
epidemic, 30, 93, 164, 170
epidemiologic, 82
epidemiology, 30, 95
esophagus, 30
ethics, 57, 212
ethnic background, 181
ethnic diversity, 185
ethnic minority, 31, 100
ethnicity, 38, 55, 64, 146, 150, 153, 158, 179
etiology, 30
Europe, 29, 142
European Commission, 66
evidence, 19, 23, 30, 55, 94, 98, 114, 126, 131, 132, 133, 134, 150, 156, 158, 187, 190, 191, 195, 196, 197, 198, 212
examinations, 143
exclusion, 191
exercise, 66, 67, 75, 78
experimental design, 196
expertise, 129
exposure, 9, 11, 15, 16, 18, 19, 106, 139, 195, 196, 197
eye movement, 79

F

factor analysis, 57, 186
faith, 177, 179
false notions, 4
families, 34, 35, 36, 37, 38, 49, 50, 89, 93, 99, 101, 103, 106, 107, 118, 128, 146
family behavior, 40, 143
family conflict, 177
family environment, 114
family functioning, 49
family history, 146
family income, 101, 156
family life, 34
family members, 4, 40, 117, 166

family system, 36
fear, 91
feelings, 10, 17, 177
fibrosis, 23, 29, 30, 31
fights, 38, 40, 41
films, 5
financial, 62, 93, 138
financial support, 138
Finland, 67, 69, 70, 82, 83
food, 83, 152
food intake, 83
force, 142
formation, 128
formula, 140
freedom, 45, 46
funds, 157

G

GCE, 84
gender differences, 15, 71, 73, 75, 150, 151
general practitioner, 68
generalized anxiety disorder, 82
Georgia, 11, 17
Germany, 67, 69, 71, 207
gerontology, 206
gingival, 23
glucose, 83
God, 179, 185
governments, 22
GPRA, 153
grades, 157
Greece, 18, 53, 56, 63, 65, 95
Greeks, 71
group membership, 177
growth, 23, 115
guardian, 90
guidelines, 87
Gulf Coast, 173

H

habitat, 26
hallucinations, 141

hazardous behavior, 5
hazards, 4, 13, 27, 28, 71, 78, 98, 164, 168, 169
head and neck cancer, 23, 27, 29, 30, 31
headache, 23
healing, 212
health care, 190, 211
health care costs, 190
health education, 63
health effects, 10, 210
health problems, 10, 177
health promotion, 107, 132, 196
health researchers, 57
health services, 186, 188
health status, 71, 101
heart disease, 4
heavy drinking, 65, 138, 177, 188
height, 175
hepatocellular carcinoma, 30
heroin, 190
high school, 3, 5, 19, 54, 92, 93, 115, 129, 140, 152, 158, 169
high school degree, 152
history, 4, 14, 16, 19, 22, 30, 176
HIV, 212
holistic medicine, 211
homeostasis, 80
homes, 34
homogeneity, 59
Hong Kong, 207, 211
hops, 129
hormones, 78, 86
hospitality, 197, 198
hostility, 40
house, 50, 149, 206
household composition, 153
household income, 37, 40, 44, 45, 46, 47, 48
housing, 103
human, xiii, 31, 81, 142, 203, 205, 208, 209
human behavior, 142
human development, xiii, 203, 205, 208, 209
human subjects, 31
hydroxide, 22
hyperactivity, 4, 5

hypersomnia, 68
hypertension, 23
hypothesis, 175

I

ideal, 134
identification, 81
identity, 150
illicit drug use, 65, 156
illicit substances, 64, 164
image, 99
immigrants, 22, 29, 36
immunodeficiency, 212
imprisonment, 146
incidence, 17, 23, 51, 68, 71, 73, 80, 139
income, 38, 43, 45, 46, 47, 48, 88, 99, 101, 103, 105, 106, 107, 108, 152
independence, 58, 150
independent variable, 60, 61, 76, 117, 121
India, 22, 27, 30, 31, 99, 142
Indians, 142
individual character, 115
individual characteristics, 115
individuals, 25, 31, 38, 55, 70, 79, 114, 138, 139, 176
induction, 139
industries, 151
industry, 98, 99, 105, 107, 108, 151, 170
infants, 130
informed consent, 25
ingest, 181
ingredients, 22, 23, 142
inhibition, 82
initiation, 5, 19, 35, 54, 88, 115, 127, 157, 164, 166, 167, 169, 170
injury, xii, 131, 132, 133, 134
injury prevention, 134
insomnia, 68, 69, 70, 71, 73, 74, 75, 78, 79, 80, 81, 82, 83, 190
institutional change, 177
institutions, 95, 114
insulin, 83
insulin sensitivity, 83
integration, 209

integrity, 82
intelligence, 146
interaction effect, 43
internal consistency, 179, 183
interpersonal relations, 117
interpersonal skills, 151
interrelations, 65
intervention, 54, 80, 86, 94, 132, 133, 146,
    152, 153, 168, 170, 174, 175, 179, 196,
    197
intimacy, 175
intrinsic motivation, 175
Iran, 21, 23, 27, 28, 163, 164
Ireland, 18, 195
iron, 116, 130
irritability, 141, 190
Islam, 17
Israel, 3, 137, 195, 203, 205, 206, 208, 213
issues, 63, 127, 182
Item response theory, 187

leisure, 70, 81, 166
lesions, 30, 31
lethargy, 141
level of education, 27, 152
life course, 159
lifestyle changes, 146
lifetime, 21, 24, 25, 26, 27, 28, 38, 103, 106,
    113, 115, 116, 118, 119, 139, 153, 154
light, 85, 88, 90, 93, 125, 151, 153
Likert scale, 71
liver, 140
liver cirrhosis, 140
logistic regression, 12, 25, 43, 54, 58, 60,
    62, 115, 154
longitudinal study, 65, 66, 158
love, 126, 179
low risk, 94, 187, 190
lower appetite, 4
lung cancer, 4, 164

J

Jamaica, 139
Japan, 68, 71, 82
Jordan, 15
joyriding, 39, 41
juveniles, 169

K

Korea, 18, 71, 113
Kuwait, 105, 108

L

labeling, 65
labor market, 158
laptop, 37
latency, 79
Latin America, 115
lead, 23, 76, 94, 127, 144
learning, 138, 150, 176
legislation, 146, 195, 196

M

magazines, 13
magnitude, 73, 81, 94, 115, 116, 125
Mainland China, 108
majority, 14, 50, 90, 93, 100, 101, 105, 119,
    140, 151, 152, 164, 176, 190
malaise, 190
Malaysia, 96
malnutrition, 143
man, 141
management, 24, 51, 82, 88, 132, 151, 189,
    190, 193, 211
MANOVA, 59, 101, 103, 104, 166
marijuana, 4, 51, 137, 138, 139, 140, 141,
    142, 143, 144, 145, 146, 153
marital status, 70, 72, 101, 153
marketing, 105, 108
marriage, 140, 143
mass, 15, 73, 75, 79, 108
mass media, 15
matter, 129, 164
maturation process, 146
Mauritius, 108
measurement(s), 38, 51, 94

media, 11
median, 12, 25, 181
medical, 4, 5, 29, 31, 68, 78, 109, 203, 211
medication, 190
medicine, 74, 94, 141, 163, 203, 208, 209, 212
Mediterranean, 10, 15, 95, 170
mellitus, 23
membership, 41
memory, 55
mental disorder, 80, 187, 211
mental health, 69, 81, 107, 117, 138, 158, 165, 176, 186, 188
mental state, 5
mentorship, 151
messages, 19
metabolism, 79, 83, 86
methadone, 189, 190, 191, 192
methodology, 11, 16
Mexico, 129
Middle East, 9
migration, 29, 36, 70, 143, 211
military, 109
mind-body, 212
miniature, 144
Ministry of Education, 87, 95, 163
minors, 5, 106, 152
mission, 205
misuse, 35, 51, 86, 132, 147, 171, 187
MMP, 138, 146
models, 17, 45, 46, 49, 62, 93, 113, 115, 118, 120, 125, 126, 127, 151, 174
modern society, 34
modifications, 38
mold, 210
monopoly, 99
mood disorder, 147, 211
morbidity, 21, 164
mortality, 10, 21, 23, 86, 108, 163, 164, 170, 190, 195
mucosa, 23
multiple regression, 115
multiple regression analysis, 115
multiplication, 37
multivariate analysis, 26, 27, 28, 58

murder, 108
music, 144
mycotoxins, 210
myocardial infarction, 198

N

National Survey, 52, 153
nausea, 141
negative consequences, 54, 55, 62
negative coping, 175
negative effects, 138
negative outcomes, 62
negative reinforcement, 55
neglect, 179, 210
neighborhood characteristics, 116
Netherlands, 147
New Zealand, 19
next generation, 208
nicotine, 4, 22, 57, 79, 86, 88, 94, 95, 108
Nigeria, 85, 87, 92, 95
nitrosamines, 30
Nobel Prize, 203
Norway, 69, 82, 207
nursing, 132, 190, 196
nutrition, 213
nutritional status, 116

O

obesity, 78, 79, 83
obstacles, 91
offenders, 186
opportunities, 36, 71, 98, 116, 127, 150, 151
optimism, 146, 177
oral cavity, 23, 27, 30
oral health, 31
Organization of American States, 115
outpatient, 68, 82

P

Pacific, 33, 152
pain, 143, 210

Pakistan, 22, 23, 27, 28, 29, 31
palpitations, 141
Panama, 129
parallel, 66
parental consent, 152
parental control, 35
parental influence, 49, 114
parental smoking, 10, 13, 15
parenting, 33, 34, 35, 36, 37, 39, 42, 43, 44, 45, 46, 47, 48, 49, 50, 51, 114, 127, 129
parenting styles, 42, 49, 51, 127
parents, 4, 10, 12, 13, 15, 17, 34, 36, 38, 39, 40, 42, 85, 88, 89, 90, 91, 93, 114, 116, 117, 118, 125, 127, 128, 129, 143, 152, 166
participants, 13, 14, 15, 16, 54, 56, 63, 65, 72, 132, 150, 152, 153, 154, 155, 156, 178, 180, 181, 184, 190, 192, 196, 197
path analysis, 171
pathways, 49, 65
peace, 177
peer group, 89, 98, 174
peer influence, 19, 129
peer review, 21, 24
peptic ulcer, 27, 31
peptic ulcer disease, 27, 31
perceived health, 98, 102
personal life, 139
personality, 65, 137, 146
personality factors, 146
pharynx, 27, 30
Philadelphia, 94, 95
physical activity, 4, 70, 78, 81
physical health, 68, 75, 76, 78, 80, 81
physical withdrawal, 190
physiology, 81
pilot study, 27, 29, 171
plants, 138
platform, 173, 178, 180
police, 38, 41
policy, 30, 54, 78, 157, 174, 198, 208
policy initiative, 30
policy makers, 78, 157, 198
politics, 212

population, 16, 22, 27, 30, 36, 37, 56, 57, 63, 68, 69, 70, 82, 83, 92, 94, 95, 99, 100, 105, 106, 107, 115, 128, 143, 164, 165, 173, 178, 180, 183, 185, 196, 197, 198
population group, 107, 183
Portugal, 87, 92
positive effects of smoking, 4
positive mental health, 177
positive reinforcement, 55
positive relationship, 116
precedent, 115, 185
predictability, 36
predictor variables, 12
prefrontal cortex, 82
premature death, 98, 164
preparation, 50, 152
prevalence rate, 86, 92, 166
preventable morbidity, 10, 86, 163, 164
prevention, 5, 34, 49, 50, 63, 65, 78, 80, 98, 106, 134, 152, 157, 170, 171, 174, 175, 178, 186, 187, 188, 203, 211
preventive approach, 176
preventive programs, 28, 165
primary school, 93
principles, 31, 212
probability, 11, 37, 113, 116, 118, 120, 121, 125, 126, 128, 186, 188
probability sampling, 11
problem behavior(s), 18, 35, 50, 157, 158
problem drinking, 186
problem-solving, 164
problem-solving skills, 164
procurement, 88, 93
professionals, 176, 206, 208
prognosis, 198
programming, 174, 177, 178, 182, 184, 185
project, 18, 65, 94, 128, 129, 171
protection, 34, 36, 40
protective factors, 115, 139
protective role, 114, 128
pro-tobacco advertisements, 10, 15, 16, 17
psychiatric disorders, 18
psychiatry, 211
psychological distress, 156

psychological health, 68, 71, 77, 78, 81, 107, 126
psychology, 22, 24, 187, 211
psychometric properties, 57
psychopharmacology, 211
psychosis, 138
psychotherapy, 82
public domain, 179
public health, 10, 23, 70, 86, 94, 98, 100, 132, 190, 203, 205, 206, 208, 209, 210, 212, 213
public service, 205
public support, 198
punishment, 117

Q

qualitative research, 184
quality of life, 203, 210, 213
query, 173
questionnaire, 11, 12, 21, 24, 25, 37, 51, 57, 65, 66, 70, 72, 85, 87, 88, 98, 100, 107, 108, 116, 164

R

race, 4, 150, 153, 158
radio, 90
rate of return, 180
reactivity, 82
reading, 182
reality, 177
reasoning, 40
recall, 68
reciprocity, 19
recognition, 175, 176
recommendations, 134, 192, 198
recovery, 176, 185, 187, 188, 213
recreation, 143
regression, 12, 25, 43, 45, 58, 62, 73, 75, 115, 120, 133, 154, 155
regression analysis, 12, 73, 75, 133, 155
regression model, 115
regulations, 106, 132

rehabilitation, 203, 205
reinforcers, 175
relationship quality, 52
relatives, 90
relaxation, 55
relevance, 176
reliability, 16, 51, 96, 102, 166, 179, 181, 186
relief, 143, 175
religion, 138, 176, 185, 187
religious beliefs, 34
religiousness, 188
replication, 183, 185, 186
representativeness, 63
research online, 83
researchers, 37, 64, 71, 107, 114, 150, 153, 169, 175, 176, 178, 179, 182, 184, 186, 197, 209
resolution, 193
resources, 5, 28, 99, 175, 185
response, 16, 24, 42, 72, 81, 89, 100, 101, 107, 117, 153, 179, 185
responsiveness, 35
restaurants, 197
restoration, 82
restrictions, 106, 197
restructuring, 188
retail, 151
reticulum, 86
rewards, 151
rhythm, 78
risk factors, 17, 28, 29, 30, 94, 95, 115, 139, 174
risks, 10, 17, 34, 35, 40, 54, 175
romantic relationship, 141
roots, 210
routines, 51
rules, 36, 39, 42, 50, 106, 146
rural areas, 34, 49

S

sadness, 166, 169
safety, 132
saliva, 197

sample design, 11
sample survey, 183
sanctions, 51, 139, 145
Saudi Arabia, 87, 92, 95
scaling, 81
scatter, 183
school, 5, 10, 11, 14, 16, 18, 19, 35, 38, 39,
    41, 49, 50, 54, 55, 58, 65, 85, 86, 87, 88,
    89, 90, 92, 93, 94, 95, 96, 115, 130, 138,
    140, 143, 151, 152, 157, 163, 165, 166,
    170, 171, 208
science, 24, 206, 212
scripts, 146
second hand smoke, 197
secondary school students, 35, 87, 94, 95,
    96, 108, 147
secondary schools, 35, 85, 87, 89, 93
self esteem, 175, 177
self help, 185
self-confidence, 90
self-esteem, 177
self-reports, 197
sensations, 69
services, 132, 175, 180, 182, 185
SES, 127
sex, 49, 82, 99
sexual behavior, 4, 54, 65, 144
sexual intercourse, 164
sexuality, 210
shape, 174
sibling(s), 15, 85, 88, 90, 91, 93, 129
side effects, 141, 192
signs, 29, 86, 88, 189, 190, 193
Singapore, 29
skills training, 164, 165, 168, 169, 171
slavery, 142
slaves, 142
sleep deprivation, 68, 69, 81
sleep disorders, 82
sleep disturbance, 68, 83
sleep fragmentation, 79
sleep habits, 82
sleep latency, 79
sleeping pills, 80
sleeping problems, 68, 70, 73, 78, 80

smoke exposure, xiii, 195, 196, 197, 199
smoking cessation, 63, 197, 198
social acceptance, 97
social activities, 158
social adjustment, 143
social behavior, 107
social benefits, 59, 60, 62
social class, 86, 88, 89, 97, 151
social competence, 171
social consequences, 147
social context, 35
social costs, 54, 55, 56, 58, 59, 60, 61, 62,
    63
social desirability, 64
social image, 107
social influence, 15, 19, 165, 174
social interactions, 114
social norms, 86
social problems, 139
social relations, 114
social relationships, 114
social sciences, 179
social situations, 85, 89, 90, 146
social skills, 163, 165, 166, 168, 169, 170
social skills training, 163, 165, 168, 169,
    170
social status, 4
social stress, 177
social support, 74, 80, 188
social withdrawal, 138
socialization, 35, 36, 114, 127, 128, 138,
    143, 156, 157
society, 4, 97, 105, 106, 139, 169, 206
socioeconomic status, 116, 117, 118
software, 10, 11, 18, 166
solitude, 143
solution, 73, 74, 80, 131
South Asia, 22, 30
South Korea, 18, 113
Southeast Asia, 21, 22
Spain, 147
special education, 5
specifications, 122
spending, 57, 156, 157

spirituality, 176, 177, 179, 181, 182, 184, 187, 188
Spring, 83
sputum, 86
Sri Lanka, 97, 98, 99, 100, 105, 106, 107, 108
stability, 36
standard deviation, 75
standard error, 121, 124
state, 27, 28, 34, 138, 146, 157, 195
states, 22, 24, 28, 39, 63, 79, 142
statistics, 25, 54, 59, 66, 101
stigma, 140
stomach, 23
storage, 79
stress, 4, 36, 67, 71, 75, 79, 80, 97, 98, 99, 100, 105, 107, 143, 174, 175, 178, 188
stressors, 175, 177
stroke, 4
structure, 36, 57, 146, 174
style, 34, 35, 37, 39, 42, 43, 44, 45, 46, 47, 48, 49, 79, 139
substance abuse, 126, 133, 139, 143, 147, 158, 165, 169, 173, 174, 175, 176, 177, 178, 179, 181, 182, 184, 185, 186, 187, 188
Substance Abuse and Mental Health Services Administration, 157
substance use, 3, 35, 36, 50, 51, 95, 115, 117, 129, 139, 149, 150, 151, 152, 153, 156, 157, 158, 169, 171, 174, 175, 176, 178, 179, 186, 187
SUD, 174
sugarcane, 138
suicidal behavior, 139, 147
suicide, 211
Sun, 51, 211, 213
supervision, 151
support services, 196
suppression, 192
surveillance, 18, 94
survey design, 88, 100
susceptibility, 55, 127
Sweden, 69, 207
Switzerland, 16

symptoms, 5, 68, 70, 83, 88, 93, 95, 96, 141, 142, 145, 189, 190, 193, 198, 213
syndrome, 198
synthesis, 196
Syria, 87, 92, 95

T

tactics, 99
Taiwan, 71
target, 37, 93, 156, 181, 196
target population, 156, 181, 196
taxes, 5
teachers, 127
team members, 38
technical assistance, 29, 170
techniques, 51, 101
teens, 157
teeth, 4, 22, 23, 86
temperature, 78
tension, 34
testing, 131, 132, 133, 134
test-retest reliability, 16
textbook, 211
Thailand, 33, 34, 35, 36, 37, 41, 48, 50
therapeutic process, 80
therapy, 80, 83, 94, 146, 190, 191, 192, 210
thoughts, 68, 81, 82
threats, 68
time series, 132
tobacco smoke, 17, 18
tobacco smoking, 14, 18, 19, 95, 102, 195, 196
tourism, 139
toxicity, 195
toxicology, 163
trade, 143, 145
traditions, 34, 40, 49, 146, 177
training, 151, 164, 165, 166, 169, 171, 175, 206
transformation, 177
transition to adulthood, 158
translation, 41, 100
transmission, 5, 190
transport, 132

treatment, 30, 68, 80, 82, 88, 174, 175, 176, 177, 178, 179, 182, 184, 186, 187, 188, 191, 192, 193, 211
trial, 134, 152, 190, 191, 192
Trinidad, 137, 138, 139, 143, 144, 147
Trinidad and Tobago, 143, 147
trustworthiness, 129
tuberculosis, 212
Turkey, 109

U

Ukraine, 19
uniform, 150
united, 4, 5, 11, 16, 17, 29, 33, 36, 48, 92, 97, 100, 108, 115, 129, 131, 132, 134, 145, 149, 168, 170, 173, 175, 178, 203, 207
United Kingdom (UK), 29, 30, 92, 207
United Nations, 129
United States (USA), 4, 5, 11, 16, 17, 29, 30, 33, 36, 48, 82, 97, 100, 108, 115, 129, 131, 132, 133, 134, 145, 149, 168, 173, 175, 178, 197, 203
universities, 100
urban, 31, 34, 36, 49, 55, 92, 108, 152
urban life, 34, 49
US Department of Health and Human Services, 5, 170

V

validation, 51, 66, 179, 184, 186
vandalism, 41
variables, 40, 43, 44, 45, 46, 47, 48, 55, 58, 59, 62, 75, 81, 88, 89, 101, 115, 116, 117, 118, 120, 153, 154, 181
variations, 101
variety of domains, 156
vehicles, 132
Vice President, 178
victims, 98
vocational identity, 156

W

waking, 68, 70, 86, 88, 90, 91, 94
war, 137
Washington, 94, 170, 187
web, 83, 173, 174, 178, 180, 185, 187
weight control, 55, 59, 60, 62, 63
weight gain, 78, 79, 83, 141
weight loss, 4
weight management, 55, 56, 62, 78
welfare, 132, 203, 205
well-being, 177, 187, 203
wellness, 152
West Bank, 9, 10, 11, 14, 16
West Indies, 137
wheezing, 86
withdrawal, 79, 80, 141, 142, 144, 189, 190, 191, 192, 193
withdrawal symptoms, 79, 141, 144, 193
workers, 142, 197, 198
workforce, 132, 150, 156, 157
working hours, 81
workplace, xii, 131, 133, 149, 151, 152, 156, 157, 196
World Health Organization(WHO), 10, 17, 30, 87, 94, 95, 99, 105, 108, 153, 158, 169, 170, 196
worldwide, 18, 23, 30, 98, 128, 138, 170, 195
worry, 81, 82

Y

Yemen, 87, 92, 95
yes/no, 153
yield, 169
young adults, 3, 5, 19, 63, 65, 83, 106, 107, 147
young people, 4, 11, 18, 51, 98, 99, 103, 106, 107, 108, 150, 151, 152, 156
young women, 156